GCSE Biology
The Workbook

This book is for anyone doing **GCSE Biology**.

It's full of **tricky questions**... each one designed to make you **sweat** — because that's the only way you'll get any **better**.

There are questions to see **what facts** you know. There are questions to see how well you can **apply those facts**. And there are questions to see what you know about **how science works**.

It's also got some daft bits in to try and make the whole experience at least vaguely entertaining for you.

What CGP is all about

Our sole aim here at CGP is to produce the highest quality books — carefully written, immaculately presented and dangerously close to being funny.

Then we work our socks off to get them out to you — at the cheapest possible prices.

Published by Coordination Group Publications Ltd.

Editors:
Amy Boutal, Ellen Bowness, Tom Cain, Katherine Craig, Kate Houghton, Sarah Hilton, Sharon Keeley, Rose Parkin, Kate Redmond, Katherine Reed, Rachel Selway, Laurence Stamford, Ami Snelling.

Contributors:
Bridie Begbie, Steve Buckley, Claire Charlton, Jane Davies, Max Fishel, James Foster, Paddy Gannon, Dr Iona M J Hamilton, Derek Harvey, Rebecca Harvey, Barbara Mascetti, Richard Parsons, Andy Rankin, Philip Rushworth, Adrian Schmit, Claire Stebbing, Paul Warren, Anna-Fe Williamson, Dee Wyatt.

ISBN: 978 1 84146 642 2

Groovy website: www.cgpbooks.co.uk

Printed by Elanders Hindson Ltd, Newcastle upon Tyne.
Jolly bits of clipart from CorelDRAW®

The Nervous System

Q1 Complete the following passage by choosing the correct words from the box.

organs ~~motor~~ ~~effectors~~ ~~neurones~~ ~~sensory~~ ~~glands~~ ~~electrical~~

Nerve cells (neurones) transmit electrical impulses from our sense organs to the CNS. Messages from the CNS are sent to effectors, which are muscles or glands. The impulses are carried along sensory and motor neurones.

Q2 In each sentence below, underline the **sense organ** involved and write down the **type of receptor** that is detecting the stimulus.

a) Tariq puts a piece of lemon on his tongue. The lemon tastes sour.

chemical receptor

b) Siobhan wrinkles her nose as she smells something unpleasant in her baby brother's nappy.

chemical receptor

c) Lindsey covers her eyes when she sees the man in the mask jump out during a scary film.

light receptor

d) Xabi's ears were filled with the sound of the crowd cheering his outstanding goal.

hearing receptor

Q3 Some parts of the body are known as the **CNS**.

a) What do the letters **CNS** stand for? Central Nervous System

b) Name the two main parts of the CNS.

1. Brain 2. ....

c) What type of neurone:

i) carries information to the CNS? sensory

ii) carries instructions from the CNS? Motor

Q4 Explain why a man with a damaged **spinal cord** may **not** be able to feel someone touching his toe.

The toe can't finish its normal path through its spinal cord to the brain.

Reflexes

Q1 Why is a **reflex** reaction faster than a **voluntary** reaction?

receptor

The reflex reaction is faster than the voluntary because reflex you just do it whereas in voluntary you have to think.

Q2 Explain what a **reflex arc** is.

The reflex arc is a nerve cell which ~~carrie~~ carries information from the receptor to the effector.

Q3 When you touch something hot with a finger you **automatically** pull the finger away. The diagram shows some parts of the nervous system involved in this **reflex action**.

a) What type of neurone is:

i) neurone **X**? sensory

ii) neurone **Y**? relay

iii) neurone **Z**? motor

b) In what form is the information carried:

i) along neurone **X**?

electrical

ii) from neurone **X** to neurone **Y**?

chemical

c) Complete the sentence.

In this reflex action the muscle acts as the effector.

d) i) What are the gaps marked **W** on the diagram called?

ii) Explain how the impulse gets across these gaps.

The impulse gets accross the gap by the chemicals, which are released when the impulse arrives at one side of the gap.

Top Tips: Reflexes are really fast — that's the whole point of them. And the fewer synapses the signals have to cross, the faster the reaction. Doctors test people's reflexes by tapping below their knees to make their legs jerk. This reflex takes less than 50 milliseconds as only two synapses are involved.

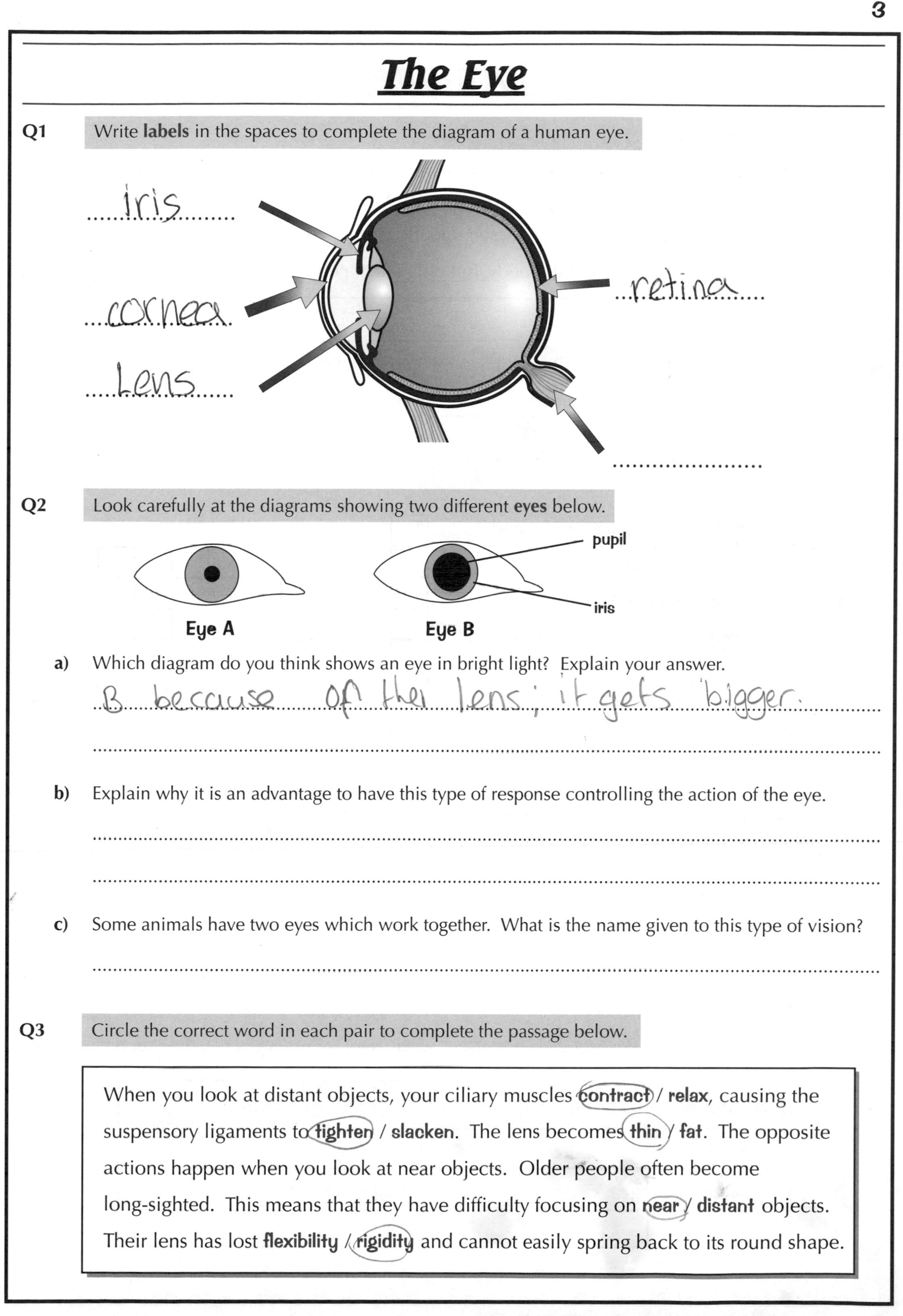

The Eye

Q1 Write **labels** in the spaces to complete the diagram of a human eye.

Q2 Look carefully at the diagrams showing two different **eyes** below.

a) Which diagram do you think shows an eye in bright light? Explain your answer.

B because of thei lens; it gets bigger.

b) Explain why it is an advantage to have this type of response controlling the action of the eye.

c) Some animals have two eyes which work together. What is the name given to this type of vision?

Q3 Circle the correct word in each pair to complete the passage below.

When you look at distant objects, your ciliary muscles **contract** / **relax**, causing the suspensory ligaments to **tighten** / **slacken**. The lens becomes **thin** / **fat**. The opposite actions happen when you look at near objects. Older people often become long-sighted. This means that they have difficulty focusing on **near** / **distant** objects. Their lens has lost **flexibility** / **rigidity** and cannot easily spring back to its round shape.

Hormones

Q1 Complete the passage below about **hormones**.

Hormones are .... messengers. They are produced in .... and released into the ....
They are carried all around the body, but only affect certain .... cells.

Q2 What is the '**fight or flight**' hormone? Why is it given this name?

....

....

Q3 Fit the answers to the clues into the grid.

a) Gland that produces insulin
b) Hormone produced by the pituitary
c) Insulin controls the level of this in the blood
d) Transports hormones around the body
e) Hormone produced by the testes

Q4 Describe the major differences between responses brought about by **hormones** and those due to the **nervous system**.

....

....

....

Q5 Complete the table below showing **hormones**, where they are **produced** and the **action** they have.

HORMONE	SITE OF PRODUCTION	ACTION
ADH		
Oestrogen		Helps control the menstrual cycle
Testosterone		

Puberty and the Menstrual Cycle

Q1 Describe **two** secondary sexual characteristics in:

a) **males**

1. .... 2. ....

b) **females**

1. .... 2. ....

Q2 The **levels** of hormones change during the menstrual cycle as the diagram below shows.

a) Identify the four hormones A-D:

A ....

B ....

C ....

D ....

A
B
C
D
Day 1
Day 4
Day 14
Day 28

b) Draw a line on the diagram to show the changes in the **thickness** of the **uterus lining** over this time.

c) Indicate on the diagram when a period takes place.

Q3 There are four main **hormones** involved in the menstrual cycle.

a) Complete the table to show **where** in the body each hormone is produced.

b) Give three effects that **oestrogen** has in the body of an adult woman.

HORMONE	WHERE IT IS PRODUCED
FSH	
oestrogen	
LH	
progesterone	

1. ....

2. ....

3. ....

Q4 An **egg** is usually released on day 14 of the menstrual cycle.

a) Why does the uterus wall become thick and spongy before the egg is released?

....

....

b) What happens in the uterus if the egg is not fertilised?

....

Controlling Fertility

Q1 Hormones can be used to **increase fertility**.

a) Name the hormone often given to women who aren't releasing any eggs. ....

b) The passage below explains how this hormone increases fertility.
Use the words in the box to fill in the gaps. Each word should be used once.

pituitary gland	LH	egg	FSH	ovary	oestrogen

.... stimulates the ovaries to produce ....,

which stimulates the .... to produce ....

This stimulates the .... to release an ....

Q2 Using **hormones** to increase or reduce fertility in women has some **disadvantages**.
Complete the table below to show some of the disadvantages of taking hormones.

Use	Possible disadvantages
Reducing fertility	1. .... 2. ....
Increasing fertility	1. .... 2. ....

Q3 **The pill** is an **oral contraceptive** that contains oestrogen. Explain how it is used to reduce fertility.

....

....

Q4 **In vitro fertilisation** (IVF) can help couples to have children.

a) Explain how **in vitro fertilisation** works.

....

....

....

b) Give one **advantage** and one **disadvantage** of in vitro fertilisation.

....

....

....

Homeostasis

Q1 Define **homeostasis**.

....

....

Q2 Which of the following is an example of the mechanism of **negative feedback**? Tick the correct sentence.

- ☐ **If your blood glucose level is too high, insulin is secreted, which causes glucose to be removed from the blood.**
- ☐ **If your blood increases in concentration, your brain sends signals to the liver.**
- ☐ **When you sweat, salt is lost.**
- ☐ **Ions are taken into the body in food.**

Q3 Circle the correct word(s) to complete the paragraph below.

On a **cold** / **hot** day or when you're exercising, you **sweat a lot** / **don't sweat much**, so you will produce **more** / **less** urine. The urine will be a **pale** / **dark** colour as it contains **less** / **more** water than usual. We say that the urine is more **concentrated** / **dilute** than usual.

Q4 The Big Brother contestants are getting on my nerves, so I put each of them on a treadmill and turn the setting to high (just to keep them quiet for a bit).

Will the contestants lose **more** or **less** water through the following body parts than they would if they sat still? Explain your answers.

a) Skin ....

....

....

b) Lungs ....

....

....

c) Kidneys ....

....

....

Homeostasis

Q5 The **human body** is usually maintained at a temperature of about **37 °C**.

a) Why do humans suffer ill effects if their body temperature varies too much from this temperature?

....

....

b) Which part of your body monitors your body temperature to ensure that it is kept constant?

....

c) How does your body cool down when it is too hot?

....

....

Q6 Ronald eats a meal that is very high in **salt**. Which of the answers below explain correctly how Ronald's body gets rid of the excess salt? Tick one or more boxes.

- ☐ Ronald's liver removes salt from his blood.
- ☐ Ronald loses salt in his sweat.
- ☐ Ronald's kidneys remove salt from his blood.
- ☐ Ronald's saliva becomes more salty, and the salt is lost when he breathes.
- ☐ Ronald gets rid of salt in his urine.

Q7 Mrs Finnegan had the **concentration of ions** in her urine measured on two days.

Date	6th December	20th July
Average air temperature	8 °C	24 °C
Ion concentration in urine	1.5 mg/cm^3	2.1 mg/cm^3

Assuming Mrs Finnegan always consumes exactly the same food and drink every day, suggest a reason for the different ion concentrations in her urine.

....

....

....

Top Tips: Homeostasis is a collective term for lots of different processes going on in lots of different parts of the body. What they all have in common is that they're trying to keep various things as constant and steady as possible — be it temperature, water content, blood sugar, levels of an ion...

Controlling Blood Sugar

Q1 Most people's **blood sugar** levels are controlled by **homeostasis**.

a) Where does the sugar in your blood come from?

....

b) Name the **two** main organs that are involved in the control of blood sugar levels.

....

c) Name the main hormone involved in the regulation of blood sugar levels.

....

Q2 Approximately **1.8 million** people in the UK have **diabetes**.

a) Explain what type 1 diabetes is.

....

b) How can diabetics **monitor** their blood sugar levels?

....

Q3 Ruby and Paul both have diabetes, so they **monitor** and **control** their glucose levels carefully.

a) Describe two ways that diabetics can **control** their blood sugar levels.

1. ....

2. ....

b) Ruby injects insulin just before she is about to eat a big meal. However, she has to go out at short notice and doesn't get time to eat. A few hours later, Ruby faints. Explain why this happens.

....

....

....

c) Paul goes out for dinner. He has forgotten to inject any insulin, and eats a large meal. A few hours after the meal Paul collapses and has to be taken to hospital for treatment.

i) Explain why Paul collapsed.

....

....

ii) What treatment would you expect Paul to be given when he arrives at hospital?

....

Eating Healthily

Q1 Complete the table to show the **functions** of some essential **nutrients** in our diet.

Nutrient	Function
Carbohydrates	
	Growth and repair of tissues.
	Provide energy and act as an energy store.

Q2 Draw lines to join each **nutrient** with what it is made from.

Carbohydrate

Protein

Fat

Amino acids

Glycerol

Simple sugars

Fatty acids

Q3 a) What are **essential amino acids**?

....

b) Where can you get them from?

....

Q4 Wendy and June both work as IT technicians. Wendy is training to run a marathon and goes to the gym every evening. June doesn't enjoy sport and prefers playing chess in her spare time. Who needs more **protein** and **carbohydrate** in their diet — June or Wendy? Explain your answer.

....

....

....

Q5 Different people need to eat **different amounts** of food because they have different energy requirements.

Calories are a measure of the amount of energy in food.

It is recommended that the average woman eats about 2000 calories per day, while the average man should eat about 2500 calories. Explain why there is a difference.

....

....

Diet Problems

Q1 Fifty men and fifty women were asked whether they thought they were **obese**. Each was then given a medical examination to **check** whether their own assessment of their weight was correct.

a) What percentage of people in the survey knew they were obese?

....

	Knew they were obese	Were obese but didn't know
Number of men	5	11
Number of women	6	5

b) From these results, do you think women or men are more aware about their weight? Explain why.

....

....

c) **Name** a health problem that has been linked to obesity ....

Q2 **Kwashiorkor** is a condition caused by lack of protein in the diet.

a) Why is this condition most common in poorer developing countries?

....

....

b) Calculate the RDA of protein for a 75 kg man. ....

Q3 **Anorexia nervosa** and **bulimia nervosa** are psychological disorders that can cause **under-nutrition**.

a) Explain the difference between these two disorders.

....

....

b) Give **three** health problems that these disorders may cause.

....

Q4 **Body mass index (BMI)** is calculated to help determine a person's ideal weight.

a) Daniel is 1.7 metres tall and weighs 76 kg. Calculate his BMI.

$$\text{BMI} = \frac{\text{body mass in kg}}{(\text{height in m})^2}$$

....

b) Look at the table below. What is Daniel's weight description? ....

Body Mass Index	Weight Description
below 18.5	underweight
18.5 - 24.9	normal
25 - 29.9	overweight
30 - 40	moderately obese
above 40	severely obese

c) Daniel is a professional athlete. Explain how this may affect your interpretation of his ideal weight.

....

....

....

Cholesterol and Salt

Q1 There are several **risk factors** for heart disease.

a) What is meant by a 'risk factor' for heart disease?

....

b) How does a high level of salt in the diet affect some people?

....

Q2 It is recommended that adults should eat no more than **6 g** of salt each day.

a) Your friend tells you that there is no way that she can be eating too much salt as she never sprinkles any on her food. Is she right?

....

....

b) The salt in food is usually listed as sodium in the nutritional information on the label. You can work out the amount of salt using the formula: **salt = sodium × 2.5**

It says there is **0.5 g of sodium per serving** of soup. How much salt is this?

....

Q3 This question is about **lipoproteins**.

a) What are lipoproteins?

....

b) Name the two types of lipoprotein involved in the transport of cholesterol in the blood.

....

Q4 In a minor **heart attack** the flow of blood in the heart muscle is reduced.

a) Explain how too much cholesterol can lead to this type of heart attack.

....

....

....

b) Which organ controls the level of cholesterol in the body?

....

Health Claims

Q1 Two reports on **low-fat foods** were published on one day. **Report A** appeared in a tabloid paper. It said that the manufacturers of 'Crunchy Bites' have shown that the latest girl band, Kandyfloss, lost weight using their product. **Report B** appeared in a journal and reported how 6000 volunteers lost weight during a trial of an experimental drug.

Which of these reports is likely to be the most reliable and why?

....

....

Q2 Three **weight loss methods** appeared in the headlines last week.

(1) **Hollywood star swears carrot soup aids weight loss**

(2) **Survey of 10 000 dieters shows it's exercise that counts**

(3) **Atkins works! 5000 in study lose weight... but what about their health?**

a) Which of these headlines are more likely to refer to **scientific studies**? Explain your answer.

....

....

b) Why might following the latest celebrity diet not always help you lose weight?

....

....

Q3 Scientists are still **not sure** whether there is a link between using cannabis and developing mental health problems, despite the fact that lots of studies have been carried out. Explain why this is.

....

....

Q4 A drug trial involved 6000 patients with **high cholesterol levels**. 3000 patients were given a **statin**, and 3000 were not. Both were advised to make lifestyle changes to lower their cholesterol. The decrease in their cholesterol levels compared to their levels at the start is shown on the graph.

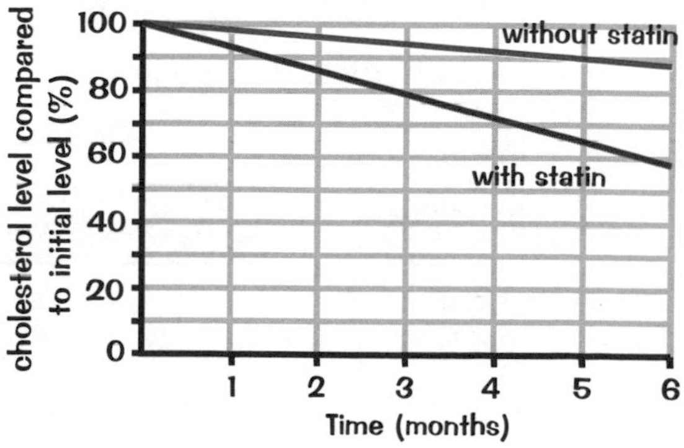

cholesterol level compared to initial level (%)
100
80
60
40
20
0
without statin
with statin
1 2 3 4 5 6
Time (months)

a) Why was the group without statin included?

....

b) Suggest a conclusion that can be drawn from these results. ....

....

Drugs

Q1 a) What is a **drug**?

....

b) What does the term **physical addiction** mean?

....

....

c) Some drugs cause the body to develop a **tolerance** to them. What does this mean?

....

....

Q2 Three people were arrested and charged with drug offences. Janice had been found smoking **cannabis** (a class C drug). Paul and Duncan had been discovered taking **amphetamine** (a class B drug). Paul had obtained the drugs and had sold some to Duncan.

a) Who would receive the **most** severe punishment? Explain your answer.

....

....

b) Who would receive the **least** severe punishment? Explain your answer.

....

....

Q3 Choose the correct words to complete the paragraph below.

solvents depressants decrease reactions judgement stimulants nicotine increasing ecstasy alcohol

Sedatives (also known as ....) include substances such as .... and .... They .... brain activity, causing slow .... and impaired .... (e.g. .... and ....) have the opposite effect, ... the activity of the brain.

Q4 Some people take drugs to '**lose their inhibitions**'. However, this could also mean that they take more **risks**. Give two possible examples.

1. ....

2. ....

Drug Testing

Q1 Write numbers in the boxes below to show the correct **order** in which drugs are usually developed and tested.

☐ Drug is tested on human tissue.

☐ Computer models simulate a response to the drug.

☐ Human volunteers are used to test the drug.

☐ Drug is tested on live animals.

Q2 Before drugs are made freely available, **clinical trials** must be performed.

a) Give two reasons why clinical trials have to take place before new drugs are made freely available.

1. ....

2. ....

b) Explain why clinical trials can't be done on human tissue samples only.

....

....

Q3 A pharmaceutical company is trialling a new drug. They are using a **placebo** in the trial and are conducting the trials 'double blind'.

a) What is a placebo and why is it used?

....

....

....

b) What is a double blind trial?

....

....

Q4 **Thalidomide** is a drug that was developed in the 1950s.

a) What was this drug originally developed for? ....

b) Thalidomide was not fully tested. What effect did it have when given to pregnant women?

....

....

c) Why has this drug been reintroduced recently?

....

Smoking and Alcohol

Q1 In the UK, the legal limit for alcohol in the blood when driving is **80 mg per 100 cm³**. The table shows the number of 'units' of alcohol in different drinks. One **unit** increases the blood alcohol level by over **20 mg per 100 cm³** in most people.

DRINK	ALCOHOL UNITS
1 pint of strong lager	3
1 pint of beer	2
1 single measure of whisky	1

a) Bill drinks two pints of strong lager. How many units of alcohol has he had? ....

b) Can Bill legally drive after these two pints? Explain your answer.

....

....

Assume he drank the pints fairly quickly.

c) Explain why it can be dangerous to drive a car after drinking alcohol.

....

Q2 The graph shows how the **percentage** of people aged between 35 and 54 who **smoke** has changed since 1950 in the UK.

a) Describe **two** of the main trends you can see in the graph.

....

....

....

....

....

b) Why are smokers more likely to suffer from:

i) chest infections ....

....

ii) cancers ....

Q3 Tobacco and alcohol are both **legal** drugs.

a) Why do alcohol and smoking have a **bigger impact** than illegal drugs in the UK?

....

b) Give two ways in which smoking and alcohol consumption **negatively** affect the UK **economy**.

....

....

Top Tips: Alcohol and tobacco might be legal but they can still harm the body. Alcohol can damage the liver and the brain. Cigarette smoke can cause problems like emphysema and lung cancer.

Solvents and Painkillers

Q1 Choose the correct words from the list to complete the paragraph below.

brain damage speeding up irritate slowing down relationship stimulants depressants respiratory weight gain soothe

Like alcohol, solvents are .... Solvents affect the nervous system by .... messages as they're passed along neurones. Long term solvent abuse can cause .... In the short term, solvents can .... the lungs and breathing passages causing .... problems.

Q2 After a night of heavy drinking Helen wakes up with a headache so she decides to take some **paracetamol**. The instructions on the bottle say that **two** tablets should be taken every **four** hours. As it is such a bad headache she decides to take a couple of extra tablets.

a) Give two reasons why this is not a good idea.

1. ....

2. ....

b) What symptoms can be treated using paracetamol?

....

Q3 **Opiates** are a group of **painkilling** drugs that come from the opium poppy.

a) Circle the drugs below that are opiates.

cannabis opium ibuprofen morphine paracetamol heroin codeine amphetamines

b) Why are some opiates only ever administered under medical supervision?

....

Q4 Different **painkillers** work in different ways.

a) How do opiates work?

....

b) Aspirin works by inhibiting the formation of prostaglandins. What are prostaglandins?

....

Causes of Disease

Q1 Fill in the gaps in the passage below using the words in the box.

cells	bursts	celled	genetic material	damaging	toxins	damage	copies

Bacteria are single- .... organisms which can multiply rapidly. Some can make you ill by .... your body cells or producing .... Viruses are tiny particles — they are not .... They are often made up of a coat of protein and some .... Viruses replicate by fooling body cells into making .... of them. The cell usually then .... and releases the new virus particles. This cell .... makes you feel ill.

Q2 **Infectious** diseases are diseases which are **transmitted** from one person to another.

a) What is a pathogen?

....

b) Name **four** types of pathogen.

....

c) All pathogens are parasites. What does this mean?

....

Q3 Draw lines to link the **health problems** to their descriptions.

Diabetes	Caused by a vitamin C deficiency
Scurvy	Caused by an iron deficiency
Anaemia	Caused by lack of insulin production
Haemophilia	A genetic disorder

Q4 **Cancer** is caused by cells dividing out of control.

a) Explain the difference between **benign** and **malignant** tumours.

....

....

b) Describe two ways in which people can change their lifestyle to reduce their risk of cancer.

1. ....

2. ....

The Body's Defence Systems

Q1 The body's first line of defence against pathogens consists of both **physical** and **chemical** barriers.

a) Give an example of a chemical barrier and describe the role it plays in defending the body.

....

....

b) Give two examples of physical barriers and describe the role each plays in defending the body.

1. ....

....

2. ....

....

Q2 The body's second line of defence is **non-specific**.

a) What does this mean?

....

b) Which type of non-specific white blood cell engulfs foreign objects?

....

Q3 Explain how **inflammatory** responses help to fight infection.

....

....

Q4 One of the roles of **specific white blood cells** is to protect the body from infection.

a) Name **one** thing specific white blood cells can produce to help them carry out their roles.

....

b) The immune system is said to have a 'memory'. Explain what this means and why it is useful.

....

....

....

Top Tips: The body has various lines of defence against pathogens: chemical and physical barriers, non-specific white blood cells and specific white blood cells. Make sure you learn all the facts about each one. Guess it's not only the immune system that needs a good memory...

Vaccinations

Q1 **Immunisation** involves injecting dead or inactive microorganisms into the body.

a) Tick the correct boxes to say whether the statements about polio immunisation are **true** or **false**.

	True	False
i) The inactive polio microorganisms have some of the same antigens as the live pathogen.	☐	☐
ii) White blood cells produce antibodies against the antigens on the injected polio microorganisms.	☐	☐
iii) After polio immunisation the white bloods cells can produce antibodies to fight typhoid.	☐	☐
iv) Immunisation is a type of passive immunity.	☐	☐

b) Circle the correct word(s) from each pair to complete the paragraph.

In active immunity the immune system makes its own **antibodies** / **antigens**, but in passive immunity they come from **vaccination** / **another organism**. Active immunity is usually **permanent** / **temporary**, but passive immunity is **permanent** / **temporary**.

c) Why are dead or inactive microorganisms used in vaccinations?

....

Q2 Answer these questions about **immunisation**.

a) Describe how immunisations have changed the pattern of disease in the UK.

....

....

b) Name a disease that has been **eradicated** worldwide because of immunisation programmes.

....

c) Describe **two** problems that occasionally occur with vaccines.

....

....

Q3 The MMR vaccine protects against **measles, mumps** and **rubella**. There is a small risk of serious **side effects** to the vaccine such as meningitis or convulsions. However, the Government recommends that **all** children are given the MMR vaccine. Explain why this is.

....

....

....

Treating Disease — Past and Future

Q1 Ignaz Semmelweiss worked in a hospital in Vienna in the 1840s. The graph shows the percentage of women dying after childbirth, before and after a **change** that he made.

What was the change he made and why did it help?

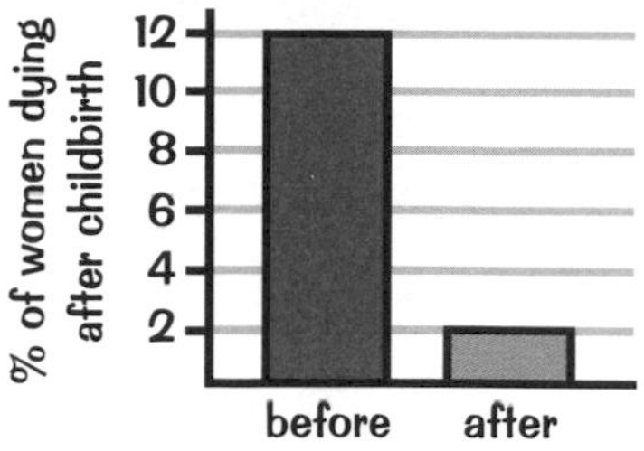

% of women dying after childbirth
12
10
8
6
4
2
before
after

Q2 Terry went to see the doctor because he had a cold. However, the doctor **wouldn't** give him any **antibiotics**.

a) Why shouldn't doctors give antibiotics for colds?

b) Why is it difficult to develop drugs to cure colds?

Q3 Jay is given **antibiotics** for an infection. Soon he feels better, so he doesn't finish the full course of antibiotics. How may this lead to the development of **antibiotic-resistant strains** of bacteria?

Q4 Some bacteria and viruses **evolve quickly**.

a) Why is rapid **bacterial** evolution a threat to human health?

b) It can be difficult to find effective **vaccines** against diseases caused by pathogens that evolve rapidly. Explain why.

Top Tips: Darwin was right when he said evolution happens gradually over many generations. The trouble is, with bacteria and viruses a whole generation can be produced in about ten minutes.

Adaptation

Q1 The picture shows two different types of **fox**.

Fox A

Fox B

a) State two differences in the appearance of the foxes.

1. ....

2. ....

b) Identify which fox lives in a cold Arctic region and which lives in a desert.

i) Fox A .... **ii)** Fox B ....

c) Explain how the features you described in part a) help each fox to survive in its natural habitat.

1. ....

....

2. ....

....

Q2 Complete the passage using some of the words given.

heat	quiet	concentrated	sweat	water	large	offspring	small	nocturnal	dilute

Mammals living in deserts need to conserve .... They make .... amounts of very .... urine. They also produce very little .... They keep cool in other ways, e.g. by being ....

Q3 The picture shows a **cactus** plant.

a) Where are cactus plants usually found? Underline the correct answer below.

In Arctic regions **In the desert** **In the mountains** **Near the sea**

b) Explain how each of the following parts of the cactus help it to survive in its normal habitat.

i) Spines ....

....

ii) Stem ....

....

iii) Roots ....

....

Classification

Q1 All organisms can be **classified** into groups, based on their similarities and differences.

a) Add the labels **genus**, **kingdom** and **species** to the diagram below.

i) ....

ii) ....

iii) ....

b) Give a definition for the term '**species**'.

....

Q2 Put ticks in the right columns to say which characteristics refer to which **kingdoms**.

	Plant	Animal
Travels to new places		
Hunts for food		
Fixed to the ground		
Compact body		

Q3 a) What is the key difference between a **vertebrate** and an **invertebrate**?

....

b) Draw lines to match the animals below to their class of vertebrate and to their descriptions.

Frog	Mammal	Feathers, most fly, mouth adapted into beak
Horse	Bird	Dry, scaly skin, lay leathery eggs on land
Snake	Fish	Moist, permeable skin, lay eggs in water
Sparrow	Amphibian	Scales and fins, live and lay eggs in water
Herring	Reptile	Produce live young, produce milk

Q4 When scientists discovered the fossilised remains of the prehistoric **archaeopteryx**, they had some difficulty in classifying it. The animal had a structure that suggested wings and feathers, but also a long bony tail, clawed hands and sharp teeth.

a) Why did scientists have difficulty in classifying the archaeopteryx?

....

b) Which features of the archaeopteryx could be described as reptilian?

....

c) Do you think the archaeopteryx laid eggs or produced live young? Explain your choice.

....

Ecosystems and Species

Q1 Here are some straightforward questions about **ecosystems** to get you started.

a) What is the difference between a natural ecosystem and an artificial ecosystem?

....

....

b) Give an example of an **artificial** ecosystem. ....

c) Which have less biodiversity — natural or artificial ecosystems? Give a reason for your answer.

....

Q2 A Year 11 class wanted to estimate the size of the **population** of clover plants on their school field.

a) If Hannah counts 11 clover plants in 1 m^2, and the school field is 250 m long by 180 m wide, how many clover plants are there likely to be on the whole field?

....

b) Hannah's friend Lisa decided to take an average from five quadrat surveys on the school field instead of only looking at one. Lisa's results are shown in the table below.

i) What is the average number of clover plants found by Lisa?

	Quadrat 1	Quadrat 2	Quadrat 3	Quadrat 4	Quadrat 5
No. plants	11	9	8	9	7

....

ii) What population size for clover would Lisa calculate?

....

iii) Whose estimation of population size is likely to be more accurate? Explain your answer.

....

Q3 A **Manx cat** was crossed with a **Siamese cat** (both domestic cats). The offspring showed characteristics of both parents, and went on to have their own offspring. A **lion** (big cat) was crossed with a **tiger** (also a big cat) and the offspring again showed features of both parents. However, all of the offspring were infertile.

What conclusions can you draw about the species of these domestic cats compared to the species of these big cats?

....

....

....

Populations and Competition

Q1 Complete the paragraph below about **red** and **grey squirrels** using some of the words from the list.

populations species competition habitat foods resources adapted

Red and grey squirrels are similar .... They like the same kind of ...., eat the same .... and find shelter in the same places. This means that they are competing for ....
Grey squirrels have proved to be better .... to the deciduous woodlands of Britain, causing numbers of the native red squirrels to fall.

Q2 The graph shows how the size of a population of **deer** and a population of **wolves** living in the same area changed over time.

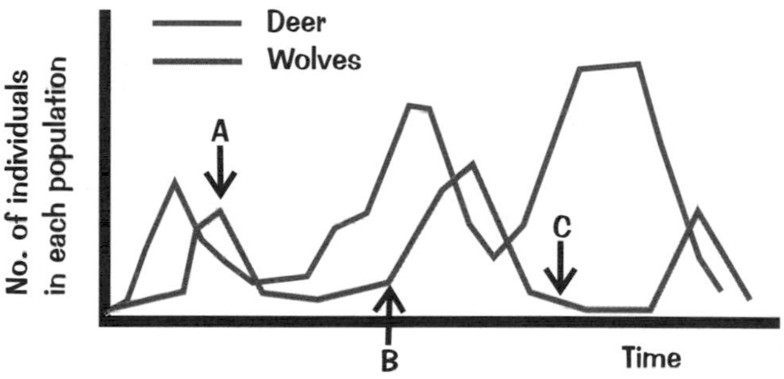

Deer
Wolves
No. of individuals in each population
A
B
C
Time

a) Describe the pattern in the changing sizes of these two populations.

....

....

b) Explain why the two populations are connected in this way.

....

c) At one point during the period covered by this graph, the wolves were affected by a disease. Underline one of the options below to show when this was.

At point A **At point B** **At point C**

d) What effect did the disease have on the size of the **deer** population? Suggest why this happened.

....

....

Q3 Some organisms, such as **parasites**, depend entirely on another species for their survival.

a) What is the difference between a **parasitic** relationship and **mutualism**?

....

....

b) State one example of mutualism. ....

Evolution

Q1 One idea of **how life began** is that simple organic molecules were brought to Earth by **comets**. It's not known if this is right.

a) What do we call this type of scientific idea? ....

b) Suggest why this idea has neither been generally accepted or completely rejected by all scientists.

....

c) Give another scientific idea for how life began.

....

....

Q2 **Fossils** are very useful to us as they show evidence of an animal or plant that lived a long time ago.

a) Choose from the words provided to complete the passage.

skeletons	habitat	decaying	colour
minerals	soft tissue	food	

Fossils form in rocks as .... replace .... tissue. They can show us features like shells, footprints, ... and occasionally .... Fossils can give us clues about the .... and .... source of an organism that lived thousands of years ago.

b) Explain why there are gaps in the fossil record.

....

....

Q3 Scientists have found remains of a **skull** which looks partly like a **chimp's** and partly like a **human's**.

a) Explain how the scientists could use this evidence to support the theory of **human evolution**.

....

....

b) How might **creationists** explain the skull?

....

....

Evolution

Q4 Explain the role of **evolution** in each of the examples below.

a) Rats are more resistant to the poison warfarin now than they were 30 years ago.

....

....

....

b) The peppered moth is black in industrial areas, but paler in rural areas.

....

....

....

c) Every few years, scientists have to create a new drug to kill the organism that causes malaria.

....

....

....

Q5 Dinosaurs, mammoths and dodos are all animals that are now **extinct.**

a) What does the term 'extinct' mean?

....

....

b) How do we know about extinct animals?

....

....

c) List **three** ways in which a species can become extinct.

....

....

....

Top Tips: We hear a lot about extinction nowadays, but it's always happened naturally. The problem today is that, due to human activity, it's happening far more often than is normal. And if you remove too many species too quickly, you unbalance things so much that the whole lot might just go...

Natural Selection

Q1 The theory of evolution by **natural selection** was developed by Charles Darwin. Tick the sentences below that describe aspects of natural selection correctly.

- [ ] All organisms face a struggle to survive.
- [ ] Organisms do not compete for basic resources.
- [ ] The best adapted animals and plants are most likely to survive.
- [ ] Some characteristics are passed on through reproduction from parent to offspring.
- [ ] Animals that have successfully adapted do not need to produce offspring.
- [ ] New species are formed through cross-breeding and not through adaptation.

Q2 Giraffes used to have much **shorter** necks than they do today. Write numbers in the boxes to show the **order** the statements should be in to explain how neck length changed.

[3] The giraffes competed for food from low branches. This food started to become scarce. Many giraffes died before they could breed.

[ ] More long-necked giraffes survived to breed, so more giraffes were born with long necks.

[ ] A giraffe was born with a longer neck than normal. The long-necked giraffe was able to reach more food.

[ ] All giraffes had short necks.

[ ] The long-necked giraffe survived to have lots of offspring that all had longer necks.

[6] All giraffes had long necks.

Q3 **Sickle cell anaemia** is a serious **genetic** disease that makes it harder for a person to carry enough oxygen in their blood. People who carry the allele that causes sickle cell anaemia are more resistant to **malaria**. Malaria is a fairly common disease in Africa that often kills.

Explain how natural selection means that sickle cell anaemia is a fairly common disease in Africa.

....

....

Q4 Which of the statements below gives a reason why some scientists did **not** at first agree with Darwin's ideas about **natural selection**? Circle the letter next to the correct statement.

A He could not explain how characteristics could be inherited.

B Characteristics that are caused by the environment can be inherited.

C They thought he was making up the evidence.

D They felt that Darwin was influenced by religious rather than scientific ideas.

Variation in Plants and Animals

Q1 Some examples of **characteristics** that **vary** between individual humans are given below. For each one, circle the correct word to show what causes the variation.

a) Eye colour **genes / environment / both**

b) Blood group **genes / environment / both**

c) Body weight **genes / environment / both**

d) Intelligence **genes / environment / both**

e) How likely you are to get heart disease **genes / environment / both**

Q2 Zoe and Anna are **identical twins**. Anna has dark hair and Zoe is blonde.

a) Do you think that both girls have kept their natural hair colour? Explain your answer.

....

....

b) Anna weighs 7 kg more than Zoe. Say whether this is due to genes, environment or both, and explain your answer.

....

....

c) Anna has a birthmark on her shoulder shaped like Wayne Rooney. Zoe doesn't. Do you think birthmarks are caused by your genes? Explain why.

....

....

Q3 When Alex looked at the ivy plant growing up the oak tree in his back garden, he was surprised by how much the **size** and **colour** of the leaves **varied**.

a) Is this variation due to **environmental** factors, **genetic** factors or **both**?

....

b) Suggest what may have affected the size and colour of the ivy leaves.

....

c) Ivy plants have a very distinctive shape to their leaves. All mature leaves have the same shape. Is this genetically or environmentally determined? ....

d) An oak tree in a different garden was planted at the same time as the one in Alex's garden. However, this oak tree is much smaller. Explain why this may be.

....

....

Genes and Chromosomes

Q1 Complete the passage using some of the words given below.

DNA	nucleus	genes	chromosomes	membrane	allele

Most cells in the body contain a structure called the ....
This contains strands of genetic information, packaged into ....
These strands are made of a chemical called .... Sections of genetic material that control different characteristics are called ....

Q2 Write out these structures in order of size, **starting with the smallest**.

nucleus	gene	chromosome	cell

1. .... 2. .... 3. .... 4. ....

Q3 Which of the following is the correct definition of the term '**alleles**'? Underline your choice.

'Alleles' is the collective term for all the genes found on a pair of chromosomes.

'Alleles' are different forms of the same gene.

'Alleles' are identical organisms produced by asexual reproduction.

Q4 Only one of the following statements is true. Tick the correct one.

There are two chromosome 7s in a human nucleus, both from the person's mother. ☐

There are two chromosome 7s in a human nucleus, both from the person's father. ☐

There are two chromosome 7s in a human nucleus, one from each parent. ☐

There is only one chromosome 7 in a human nucleus. ☐

Q5 Tick the correct boxes to show whether each statement is **true** or **false**.

		True	False
a)	Human body cells contain 44 chromosomes.	☐	☐
b)	Chromosomes are long lengths of DNA coiled up.	☐	☐
c)	Different versions of the same gene are called gametes.	☐	☐

Top Tips: First of all you need to know exactly what's meant by genes, alleles, DNA, chromosomes, etc. And don't forget that virtually all organisms have two of each chromosome in all of their normal body cells.

Sexual Reproduction and Variation

Q1 Circle the correct word(s) in each statement below to complete the sentences.

a) Sexual reproduction involves **one** / **two** individual(s).

b) The cells that are involved in sexual reproduction are called **clones** / **gametes**.

c) Sexual reproduction produces offspring with **different** / **identical** genes to the parent.

d) In sexual reproduction the sperm cell contains **the same number of** / **half as many** chromosomes as the **fertilised** egg.

Q2 Explain how a human baby receives genes from **both** its father and its mother, but still **only** has **46 chromosomes** in its cells.

....

....

....

Q3 a) What is a **mutation**?

....

b) List **four** things that can cause mutations.

....

....

c) What name is given to chemicals that can cause cancer? ....

d) How do mutations affect proteins?

....

e) Give two examples of how a mutation can be **harmful**.

1. ....

....

2. ....

....

f) Explain how a mutation can be **beneficial**.

....

....

Genetic Diagrams

Q1 **Genes** can exist in different versions called **alleles**.

a) In an individual, are both alleles always expressed? ....

If an allele is expressed, it has an effect on the organism.

b) When looking at a genetic diagram, how would you know which was a dominant allele and which was a recessive allele?

....

....

c) An organism with two of the same alleles is known as what? ....

Q2 A type of fly usually has **red** eyes. However, there are a small number of white-eyed flies. Having **white** eyes is a **recessive** characteristic.

a) Complete the following sentences with either '**red eyes**' or '**white eyes**'.

i) **R** is the allele for ....

ii) **r** is the allele for ....

iii) Flies with the alleles **RR** or **Rr** will have ....

iv) Flies with the alleles **rr** will have ....

b) Two flies have the alleles **Rr**. They fall in love and get it on.

i) Complete this genetic diagram to show the alleles of the possible offspring.

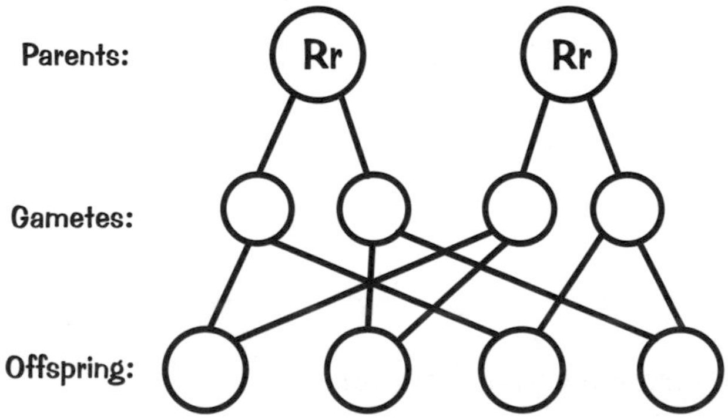

Parents: Rr Rr

Gametes:

Offspring:

ii) What is the probability that one of the flies' offspring will have white eyes?

....

iii) The flies have 96 offspring. How many of the offspring are **likely** to have red eyes?

....

Top Tips: Tackle genetic diagrams step-by-step and they'll come out OK. Remember — if the allele for a characteristic is recessive, you need two of them to show that characteristic. If you haven't got two recessive alleles, you must have at least one dominant one, so that characteristic will show up.

Genetic Disorders

Q1 Complete the passage using some of the words given below.

allele	two	carrier	genetic	parents	pancreas	three

Cystic fibrosis is a .... disorder. It is inherited from the ....
The disease is caused by a recessive .... That means a person must have .... copies of the faulty allele. They will have the symptoms of sticky mucus in the air passages and .... A person with one copy of the recessive allele is called a .... They will not have any symptoms.

Q2 The **allele** causing cystic fibrosis is **recessive**.

a) Complete the genetic diagram for the inheritance of cystic fibrosis.

b) What is the probability of a child having cystic fibrosis in this case?

....

Parents: Ff Ff

Gametes:

Offspring:

Q3 Sarah and Ahmad are expecting a baby. Tests have revealed that there is a high chance that the baby will have a **genetic disorder** that could result in death within 1 year of life. They have to decide whether to **terminate** the pregnancy.

a) Give one argument in **favour** of a termination of the pregnancy.

....

....

b) Give one argument **against** a termination of the pregnancy.

....

....

Q4 Scientists are trying to treat cystic fibrosis by inserting a **healthy copy** of the gene into airway cells.

a) What is the name given to this method of treating genetic disorders?

....

b) If this method is successful, will it prevent people with cystic fibrosis passing the faulty gene on to their children? Explain your answer.

....

....

Cloning

Q1 Complete the following sentences.

a) Offspring that are identical to their parent are called ....

b) This type of reproduction is called .... reproduction.

c) Each .... in the parent nucleus splits in half to give two identical sets of genetic information.

d) An example of an organism that reproduces in this way is a ....

Billy...?

Q2 Lucy cut her hand, but a week later she noticed that the cut had almost disappeared. The skin covering it looked just the same as the skin on the rest of her hand. This happened by the same process as **asexual reproduction**.

a) Where did the new skin cells on Lucy's hand come from?

....

....

b) Suggest why the skin on Lucy's hand looked the same as it had before she had cut herself.

....

....

c) Suggest why it took a week for the cut to heal.

....

....

Q3 a) Name the **two** methods commonly used by man to produce clones of **plants**.

1. ....

2. ....

b) Give **two** advantages and **one** disadvantage of cloning plants using methods like these.

Advantage 1: ....

Advantage 2: ....

Disadvantage: ....

Top Tips: Cloning (and genetic engineering, coming up next) are quite controversial topics. People fret about them being unnatural and having future consequences that no-one could predict. But remember that not all cloning is done by men in white coats — quite a few plants manage it too.

Cloning

Q4 Joe has a herd of cows and he wants them all to have calves, but he **only** wants to breed from his champion bull and prize cow.

a) Name a method Joe could use to achieve this.

....

b) Describe the steps involved in this method in detail.

....

....

....

c) Which of the animals involved in this process will be genetically identical?

....

d) Give one disadvantage of this method.

....

....

Q5 Read the passage before deciding whether the statements that follow are **true** or **false**.

> Sperm was collected from Gerald the stallion and used to artificially inseminate Daisy the mare. An embryo was then removed from Daisy and divided into separate cells, each of which was allowed to grow into a new embryo. These new embryos were then implanted into other horses, including Rosie, Ruby and Jilly.

		True	False
a)	Each embryo is genetically identical to Daisy.	☐	☐
b)	Gerald is genetically identical to the embryos.	☐	☐
c)	All the embryos are genetically identical.	☐	☐
d)	The embryo carried by Jilly is her natural daughter.	☐	☐
e)	All the embryos carry some of Daisy's genes.	☐	☐

Q6 Discuss the **ethical** issues involved in using **embryonic stem cells** to treat diseases.

....

....

....

....

Genetic Engineering

Q1 Fill in the gaps in the passage below to explain how **genetic engineering** is carried out.

The useful .... is 'cut' from the donor organism's chromosome using .... The same .... are then used to cut the host organism's chromosome and the useful .... is inserted.

Q2 Explain how **genetic engineering** can be used for the following:

a) Producing large amounts of human insulin in a short time.

....

....

b) Improving health in developing nations.

....

....

Q3 Some people are **worried** about genetic engineering.

a) Explain why some people are concerned about genetic engineering.

....

....

b) Do you think that scientists should be carrying out genetic engineering? Explain your answer.

....

....

....

Q4 Look carefully at this headline about a new type of **GM salmon**.

Monster food? Scientists insert a growth hormone gene and create fish that grow much faster than ever before!

Some scientists have warned that the GM salmon should be tightly controlled so they don't escape into the sea. What might happen if the GM salmon were allowed to escape?

....

....

There's Too Many People

Q1 The size of the Earth's **human population** has an impact on our environment.

a) Use the table below to plot a graph on the grid, showing how the world's human population has changed over the last 1000 years.

NO. OF PEOPLE (BILLIONS)	YEAR
0.3	1000
0.4	1200
0.4	1400
0.6	1600
1.0	1800
1.7	1900
6.1	2000

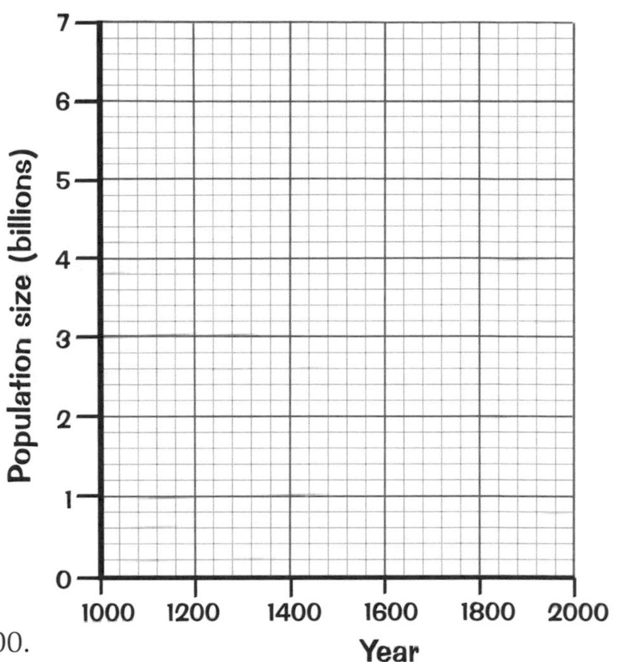

b) Suggest two reasons for the faster increase after 1800.

....

....

Q2 One way to assess a person's impact on the Earth is to use an **ecological footprint**. This involves calculating **how many Earths** would be needed if everyone lived like that person. It takes into account things like the amount of **waste** the person produces and how much **energy** they use.

a) Two men calculate their ecological footprints. Eight Earths would be needed to support everyone in the way John lives. Half an Earth would be enough to support everyone in the way Derek lives.

i) One of the men lives in a UK city, and one in rural Kenya. Who is more likely to live where?

....

ii) Tick any of the following that are possible reasons for the difference in results.

- [ ] John buys more belongings, which use more raw materials to manufacture.
- [ ] John has central heating in his home and Derek has a wood fire.
- [ ] John throws away less waste.
- [ ] John drives a car and Derek rides a bicycle.

b) Suggest one thing John could do to reduce the size of his ecological footprint.

....

Top Tips: There's lots to think about with this topic. It's the kind of thing you might get a longer answer question on in an exam, where you have to weigh up all the different arguments. And of course, examiners can't get enough of that graph where the human population suddenly increases — they love it.

The Greenhouse Effect

Q1 Underline any statements below about the greenhouse effect that are **true**.

The greenhouse effect is needed for life on Earth as we know it.

Greenhouse gases include carbon dioxide and sulfur dioxide.

The greenhouse effect causes acid rain.

Increasing amounts of greenhouse gases may lead to global warming.

Q2 The Earth receives energy from the **Sun**. It radiates much of this energy back outwards.

a) Explain the role of the greenhouse gases in keeping the Earth warm.

....

....

b) What would happen if there were no greenhouse gases?

....

c) In recent years the amounts of greenhouse gases in the atmosphere have increased.
Explain how this leads to global warming.

....

....

Q3 **Deforestation** increases the amount of **carbon dioxide** (a greenhouse gas) released into the atmosphere and decreases the amount removed.

a) Explain how this happens.

....

....

....

....

b) Give two other examples of human activities that release carbon dioxide into the atmosphere.

....

c) Name another greenhouse gas with rising levels and give two sources of this gas.

Gas: ....

Sources: ....

....

Climate Change

Q1 One UK newspaper said that **global warming** will be good for the UK because people will be able to have more barbecues. Do you think they're right? Explain your answer.

....

....

Q2 Global warming may cause the seas to warm and **expand**, putting low lying areas at increased risk from **flooding**. This isn't the only possible consequence though — fill in the flow chart to show how temperatures might **decrease**.

Higher temperatures make ice melt.
Ocean currents are disrupted.
Some areas (maybe the UK) get colder.
Cold fresh water enters the ocean.

Q3 Two university students carried out **observations**. Student A noticed that a glacier was melting. Student B noticed that daffodils flowered earlier in 2006 than in 2005. Both students concluded that their observations were due to **global warming**. Are they right? Explain your answer.

....

....

Q4 Scientists are collecting **evidence** to try to support or disprove the **theory** of global warming.

a) What is meant by evidence and theory?

evidence: ....

theory: ....

b) Give examples of the sort of data that scientists are collecting about climate change.

....

....

Air Pollution

Q1 Scientists studied the number of **whales** around the coast of Australia over a period of 15 years. They found that the number of plankton-eating whales was falling and thought that this could be due to the hole in the **ozone layer** there.

a) Why would a hole in the ozone layer affect the numbers of plankton-eating whales?

....

....

b) What might have made the hole in the ozone layer?

....

....

c) Why should the scientists wear sunscreen if they are working outdoors?

....

....

Q2 **Exhaust** fumes from cars and lorries often contain **carbon monoxide** (CO).

a) Why is CO more likely to be formed in **engines** than if the fuel was burnt in the open air?

....

....

b) Why is carbon monoxide **dangerous**?

....

Q3 Use the words and phrases below to complete the paragraph.

nitric   sulfur dioxide   the greenhouse effect   sulfuric   nitrogen oxides   acid rain

When fossil fuels are burnt carbon dioxide is produced. The main problem caused by this is .... The gas .... is also produced. This comes from sulfur impurities in the fuel. When it combines with moisture in the air .... acid is produced. This falls as acid rain. In the high temperatures inside a car engine nitrogen and oxygen from the air react together to produce .... These react with moisture to produce .... acid, which is another cause of acid rain.

Sustainable Development

Q1 Humans can affect the **environment** in lots of ways.

a) Give two ways that humans can have a negative effect on the environment.

....

b) Explain why some of the negative effects caused by humans cannot easily be reversed.

....

....

Q2 **Ecosystems** like rainforests contain many different **species**. If we destroy rainforests we risk making species extinct and **reducing biodiversity**.

a) What is meant by '**reducing biodiversity**'?

....

b) What are the implications for humans of reducing biodiversity?

....

....

....

Q3 **Mayfly larvae** and **sludge worms** are often studied to see how much **sewage** is in water.

a) What is the name for an organism used in this way? ....

Juanita recorded the number of each species in water samples taken at three different distances away from a sewage outlet. Her results are shown below.

Distance (km)	No. of mayfly larvae	No. of sludge worms
1	3	20
2	11	14
3	23	7

b) Give one thing that she would have to do to make this experiment a fair test.

....

c) What can you conclude about the two organisms from these results?

....

....

d) Suggest why sewage may decrease the number of mayfly larvae.

....

....

Conservation and Recycling

Q1 Tick the boxes to show whether the following statements are **true** or **false**.

		True	False
a)	Recycling could help to slow the increase in greenhouse gas levels.	☐	☐
b)	Paper can only be recycled a limited number of times.	☐	☐
c)	Plastics can be recycled over and over again.	☐	☐
d)	Recycling costs nothing and has huge benefits for the environment.	☐	☐
e)	Recycling causes more land to be used for landfill sites.	☐	☐

Q2 Match the following methods of **woodland conservation** to their descriptions.

coppicing

reforestation

replacement planting

replanting trees that have been cut down in the past

new trees are replanted at the same rate that others are cut down

cutting trees down to just above ground level

Q3 **Recycling** helps to conserve our natural resources.

a) What is recycling?

....

....

b) Give **three** examples of materials that can be recycled.

1. ....

2. ....

3. ....

c) Does recycling usually use more or less energy than extracting the material from scratch?

....

Q4 Describe how **conservation** measures and **recycling** help us to sustain resources for future use.

....

....

....

....

Cells

Q1 Tick the boxes to show if the statements are **true** or **false**.

		True	False
a)	Most cells are specialised for their function.	☐	☐
b)	Red blood cells, like all animal cells, have a cell wall.	☐	☐
c)	Cells are grouped together to make up tissues such as palisade tissue.	☐	☐
d)	Similar tissues are grouped together to give an organ.	☐	☐

Q2 Complete each statement below by circling the correct word(s).

a) **Plant** / **animal** cells contain chloroplasts, but **plant** / **animal** cells do not.

b) Plant cells have a **vacuole** / **cell wall**, which is made of cellulose.

c) **Both plant and animal cells** / **Only plant cells** / **Only animal cells** contain mitochondria.

d) Chloroplasts are where **respiration** / **photosynthesis** occurs, which makes **glucose** / **water**.

Q3 Below are three features of **palisade leaf cells**. Draw lines to match each feature to its function.

Lots of chloroplasts | gives a large surface area for absorbing CO_2

Tall shape | means you can pack more cells in at the top of the leaf

Thin shape | for photosynthesis

Q4 Complete the following paragraph about **guard cells** using the words below.

turgid **flaccid** **photosynthesis** **stomata**

Guard cells open and close the .... When the plant has lots of water the guard cells are .... This makes the stomata open, so gases can be exchanged for .... When the plant is short of water the guard cells become ...., making the stomata close.

Q5 State what the following cell structures **contain** or are **made of** and what their **functions** are.

a) The **nucleus** contains ....

Its function is ....

b) **Chloroplasts** contain ....

Their function is ....

c) The **cell wall** is made of ....

Its function is ....

DNA

Q1 Choose from the words below to complete the passage about the **structure of DNA**.

uracil base adenine helix cytoplasm
guanine proteins nucleotides glycine nucleus

DNA is found in the .... of cells. It is a double-stranded .... made up of lots of groups called .... Each one contains a small molecule called a .... There are four of these — ...., cytosine, ... and thymine.

Q2 Number the statements below to show the correct order of the stages in **DNA replication**.

- ☐ Cross links form between the bases of the nucleotides and the old DNA strands.
- ☐ The DNA double helix 'unzips' to form two single strands.
- ☐ The result is two molecules of DNA identical to the original molecule of DNA.
- ☐ Free-floating nucleotides join on where the bases fit.
- ☐ The new nucleotides are joined together.

Q3 The **bases** in DNA always pair up in the **same** way.

Complete the diagram below to show which **bases** will form the complementary strand of DNA.

A	C	T	G	C	A	A	T	G
......	......	......	......	......	......	......	......	......

Q4 **Genetic fingerprinting** is a way of comparing people's DNA — it's useful in forensic science. Put these stages in DNA fingerprinting into the correct order.

Compare the unique patterns of DNA.

Separate the sections of DNA.

Collect the sample for DNA testing.

Cut the DNA into small sections.

1. ....
2. ....
3. ....
4. ....

Making Proteins

Q1 Circle the correct word(s) from each pair to complete the following sentences.

a) Proteins are made up of chains of **amino acids** / **glucose**.

b) Transamination happens in the **kidneys** / **liver**.

c) Proteins are made in the cell by organelles called **chloroplasts** / **ribosomes**.

d) Each amino acid is coded for by a sequence of **three** / **four** bases.

Q2 Number the statements below to show the correct order of the stages in **protein synthesis**.

☐ Amino acids are joined together to make a polypeptide.

☐ RNA moves out of the nucleus.

☐ RNA joins with a ribosome.

☐ A molecule of RNA is made using DNA as a template.

☐ The DNA strand unzips.

Q3 Protein synthesis involves **DNA** and **RNA**.

a) Why is the information contained in DNA copied onto a strand of RNA?

....

b) Give **one** difference between DNA and RNA.

....

c) What is the name for a section of DNA that codes for a particular protein? ....

d) How do **amino acids** determine the function of a protein?

....

....

e) What happens if we don't take in the right amounts of each amino acid in our **diet**?

....

f) Explain how DNA determines the type of cells that are produced.

....

....

g) Explain the role of the **bases** in DNA in the building of a protein.

....

....

Enzymes

Q1 a) Write a definition of the word '**enzyme**'.

....

....

b) In the space provided, draw a labelled diagram to show how an enzyme's **shape** allows it to break substances down.

Q2 Tick the correct boxes to show whether the sentences are **true** or **false**.

		True	False
a)	Most enzymes are made of fat.	☐	☐
b)	The rate of most chemical reactions can be increased by raising the temperature.	☐	☐
c)	Most cells are damaged at very high temperatures.	☐	☐
d)	Each type of enzyme can speed up a lot of different reactions.	☐	☐

Q3 Stuart has a sample of an enzyme and he is trying to find out what its **optimum pH** is. Stuart tests the enzyme by **timing** how long it takes to break down a substance at different pH levels. The results of Stuart's experiment are shown in the table below.

pH	time taken for reaction in seconds
2	101
4	83
6	17
8	76
10	99
12	102

a) Draw a line graph of the results of the experiment on the grid above.

b) What is the **optimum** pH for the enzyme? ....

c) Explain why the reaction is very slow at certain pH levels.

....

....

d) Would you expect to find this enzyme in the stomach? Explain your answer.

....

Diffusion

Q1 Complete the passage below by circling the correct word in each pair.

Diffusion is the **direct** / **random** movement of particles from an area where they are at a **higher** / **lower** concentration to an area where they are at a **higher** / **lower** concentration. The rate of diffusion is faster when the concentration gradient is **bigger** / **smaller**. It is slower when there is a **large** / **small** distance over which diffusion occurs and when there is **more** / **less** surface for diffusion to take place across.

Q2 Tick the correct boxes to show whether the sentences are **true** or **false**.

		True	False
a)	Diffusion takes place in all types of substances.	☐	☐
b)	Diffusion happens more quickly when there is a larger concentration gradient.	☐	☐
c)	A larger surface area makes diffusion happen more quickly.	☐	☐
d)	A larger distance for particles to move across speeds up the rate of diffusion.	☐	☐

Q3 The first diagram below shows a **cold cup of tea** which has just had a **sugar cube** added.

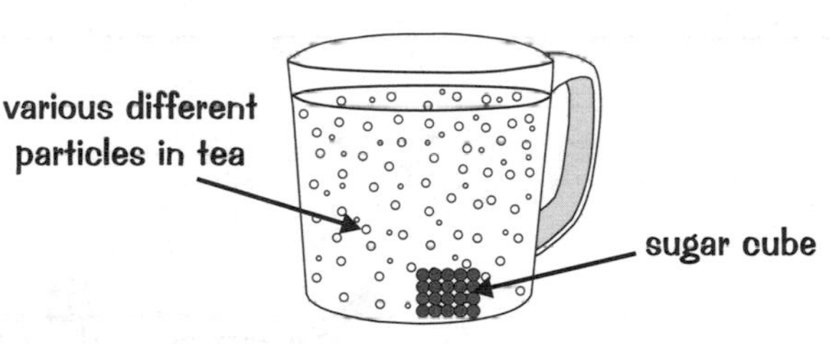

a) In the second cup above, draw the molecules of **sugar** in the tea after an hour.

b) Predict how the rate of diffusion of the sugar would change in each of the following situations:

i) sugar crystals are used rather than a sugar cube

....

ii) the tea is heated

....

iii) the sugar and tea are placed in a long thin tube

....

c) Explain the movement of the sugar particles in terms of areas of different **concentration**.

....

....

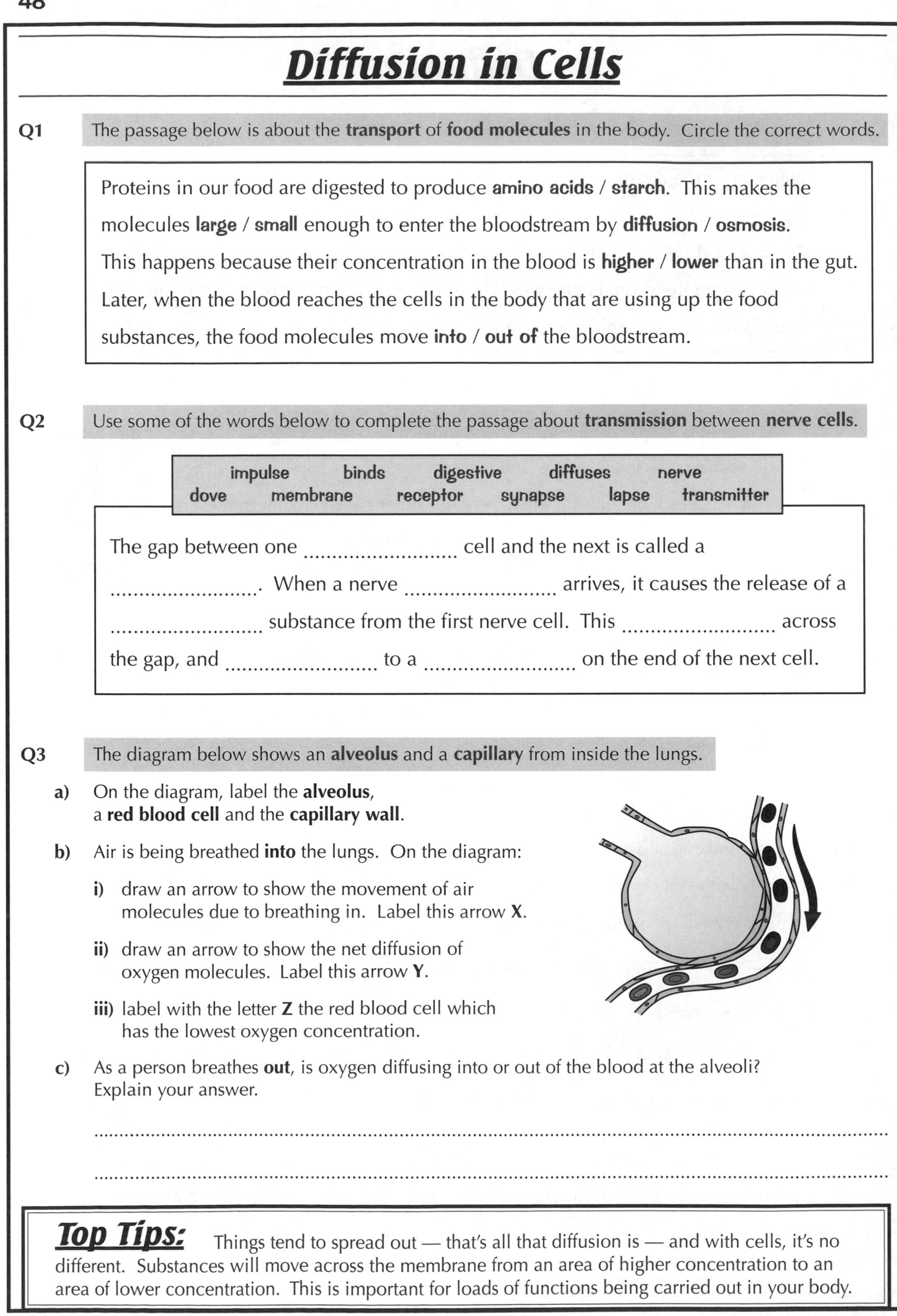

Diffusion in Cells

Q1 The passage below is about the **transport** of **food molecules** in the body. Circle the correct words.

> Proteins in our food are digested to produce **amino acids / starch**. This makes the molecules **large / small** enough to enter the bloodstream by **diffusion / osmosis.** This happens because their concentration in the blood is **higher / lower** than in the gut. Later, when the blood reaches the cells in the body that are using up the food substances, the food molecules move **into / out of** the bloodstream.

Q2 Use some of the words below to complete the passage about **transmission** between **nerve cells.**

impulse binds digestive diffuses nerve dove membrane receptor synapse lapse transmitter

> The gap between one .... cell and the next is called a .... When a nerve .... arrives, it causes the release of a .... substance from the first nerve cell. This .... across the gap, and .... to a .... on the end of the next cell.

Q3 The diagram below shows an **alveolus** and a **capillary** from inside the lungs.

a) On the diagram, label the **alveolus**, a **red blood cell** and the **capillary wall.**

b) Air is being breathed **into** the lungs. On the diagram:

i) draw an arrow to show the movement of air molecules due to breathing in. Label this arrow **X**.

ii) draw an arrow to show the net diffusion of oxygen molecules. Label this arrow **Y**.

iii) label with the letter **Z** the red blood cell which has the lowest oxygen concentration.

c) As a person breathes **out**, is oxygen diffusing into or out of the blood at the alveoli? Explain your answer.

....

....

Top Tips: Things tend to spread out — that's all that diffusion is — and with cells, it's no different. Substances will move across the membrane from an area of higher concentration to an area of lower concentration. This is important for loads of functions being carried out in your body.

Osmosis

Q1 This diagram shows a tank separated into two by a **partially permeable membrane**.

A B

Water molecule
Glucose molecule

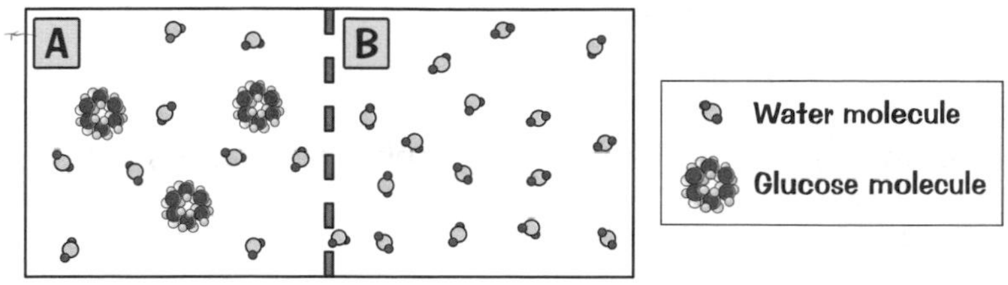

a) What is a partially permeable membrane?

....

b) On which side of the membrane is there a higher concentration of water molecules? ....

c) In which direction would you expect more water molecules to travel — from A to B or from B to A?

....

d) Predict whether the level of liquid on side B will **rise** or **fall**. Explain your answer.

The liquid level on side B will ...., because ....

....

Q2 The diagram below shows some **body cells** bathed in **tissue fluid**. A blood vessel flows close to the cells, providing water. The cells shown have a low concentration of water inside them.

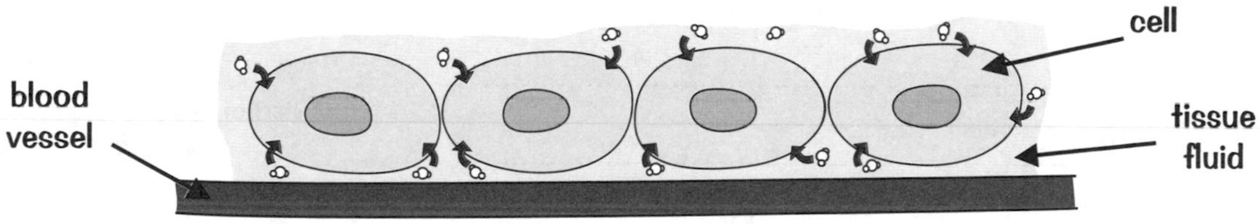

blood vessel
cell
tissue fluid

a) Is the concentration of water higher in the **tissue fluid** or inside the **cells**? ....

b) In which direction would you expect more water to travel — **into** the cells or **out of** the cells? Explain your answer.

....

....

c) Explain why osmosis appears to stop after a while.

....

....

Top Tips: Don't forget it's only small molecules that can diffuse through cell membranes, e.g. glucose, amino acids, water and oxygen. Big hulking things like proteins and starch can't fit through.

Respiration and Exercise

Q1 Tick the correct boxes to show whether the sentences are **true** or **false**.

		True	False
a)	Aerobic respiration releases energy.	☐	☐
b)	Respiration usually releases energy from protein.	☐	☐
c)	Aerobic respiration is more efficient than anaerobic respiration.	☐	☐
d)	Respiration takes place in a cell's nucleus.	☐	☐
e)	Aerobic respiration produces carbon dioxide.	☐	☐
f)	Anaerobic respiration happens when there's not enough oxygen available.	☐	☐
g)	Plants use photosynthesis instead of respiration.	☐	☐

Q2 Write the **word equations** for:

a) Aerobic respiration ....

b) Anaerobic respiration ....

Q3 Jim is a keen runner. He takes part in a 400 metre race. The **graph** below shows Jim's **breathing rate** before, during and after the race.

Breaths per minute: 0, 5, 10, 15, 20, 25, 30, 35, 40, 45, 50

Time (minutes): 1, 2, 3, 4, 5, 6

Start of race

a) How much did his breathing rate go up during the race?

.... **breaths per minute**

b) Explain why Jim's breathing rate increased.

....

....

....

c) Why doesn't Jim's breathing rate return to normal immediately after the race?

....

....

Q4 Oxygen diffuses into the blood through the walls of the **alveoli** of the lungs.

a) Describe **two** ways in which the structure of the alveoli helps them to exchange oxygen efficiently.

....

b) What effect would holding your breath have on the rate of diffusion of oxygen?

....

....

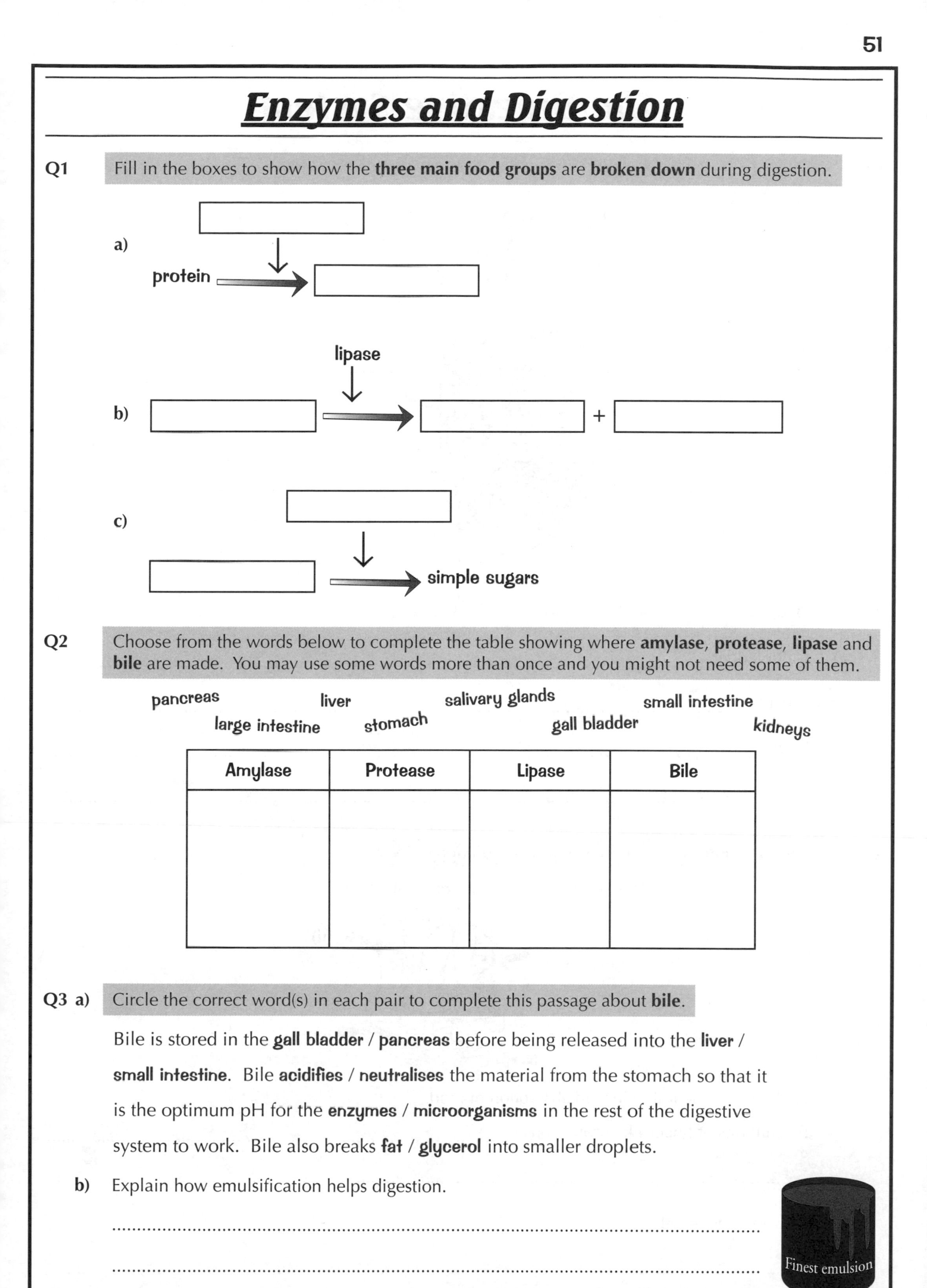

Enzymes and Digestion

Q1 Fill in the boxes to show how the **three main food groups** are **broken down** during digestion.

a) [ ] ↓ protein → [ ]

b) lipase ↓ [ ] → [ ] + [ ]

c) [ ] ↓ [ ] → simple sugars

Q2 Choose from the words below to complete the table showing where **amylase**, **protease**, **lipase** and **bile** are made. You may use some words more than once and you might not need some of them.

pancreas liver salivary glands small intestine large intestine stomach gall bladder kidneys

Amylase	Protease	Lipase	Bile

Q3 a) Circle the correct word(s) in each pair to complete this passage about **bile**.

Bile is stored in the **gall bladder** / **pancreas** before being released into the **liver** / **small intestine**. Bile **acidifies** / **neutralises** the material from the stomach so that it is the optimum pH for the **enzymes** / **microorganisms** in the rest of the digestive system to work. Bile also breaks **fat** / **glycerol** into smaller droplets.

b) Explain how emulsification helps digestion.

....

....

The Digestive System

Q1 Fill in the boxes to label this diagram of the human **digestive system**.

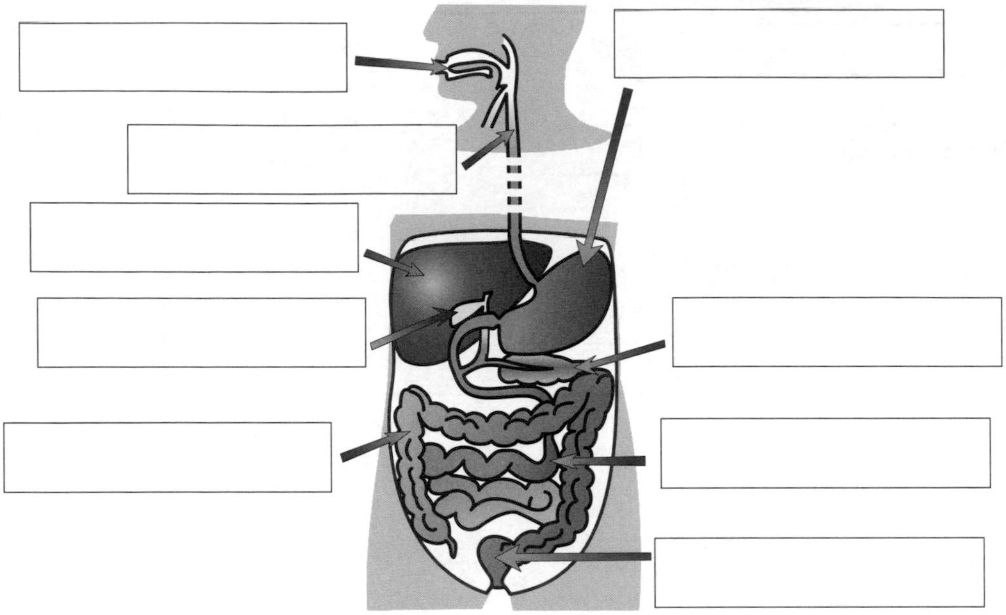

Q2 Describe the role of each of the following in **digestion**:

a) Gall bladder ....

b) Pancreas ....

c) Liver ....

d) Large intestine ....

Q3 The **small intestine** is adapted for the absorption of food.

a) Label the diagram below showing part of the **small intestine**.

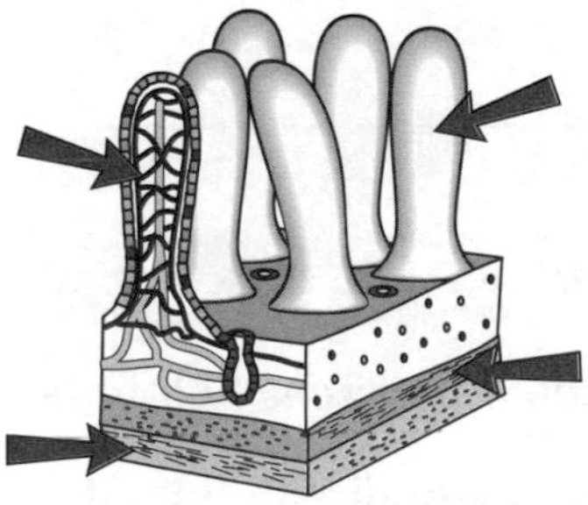

i) ....

ii) ....

iii) ....

iv) ....

b) Explain how the following aid absorption of food:

i) millions of finger-like projections ....

....

ii) very long length ....

....

Functions of the Blood

Q1 Which of these statements are **true**, and which are **false**? Tick the correct boxes.

		True	False
a)	The function of red blood cells is to fight germs.	☐	☐
b)	White blood cells help to clot blood.	☐	☐
c)	Glucose can be found in the blood.	☐	☐
d)	The liquid part of blood is called urine.	☐	☐
e)	Platelets help to seal wounds to prevent blood loss.	☐	☐

Q2 **Plasma** carries just about everything around the body.

a) For each of the substances listed in the table, state where in the body it is travelling **from** and **to**.

Substance	Travelling from	Travelling to
Urea		
Carbon dioxide		
Glucose		

b) List six other things that are carried by plasma.

1. .... 4. ....

2. .... 5. ....

3. .... 6. ....

Q3 Use some of the words below to complete the passage about the structure of **red blood cells**.

large **small** **nucleus** **flexible** **rigid**
carbon dioxide **oxygen** **cytoplasm** **haemoglobin** **oxyhaemoglobin**

Red blood cells are biconcave in shape, which means they have a .... surface area for absorbing oxygen. They have no ...., but their cytoplasm is full of ...., which can combine with .... to form .... Red blood cells are very ...., which means that they can fit easily through capillaries.

Q4 The main role of **white blood cells** is defence against disease.

a) What do they produce to fight microbes? ....

b) What do they produce to neutralise the toxins produced by microbes? ....

c) Explain how white blood cells are able to digest microorganisms.

....

....

Blood Vessels

Q1 Draw lines to match each of the words below with its correct description.

artery | hole in the middle of a tube
capillary | microscopic blood vessel
lumen | vessel that takes blood towards the heart
vein | vessel that takes blood away from the heart

Q2 Circle the correct word in each of the sentences below.

a) **Arteries** / **Veins** contain valves to prevent the blood going backwards.

b) **Capillaries** / **Veins** have walls that are only one cell thick.

c) The blood pressure in the **arteries** / **veins** is higher than in the **arteries** / **veins**.

Q3 **Cholesterol** is a fatty substance needed in the body.

a) Why do you need cholesterol in your body? ....

b) Complete the following sentence.

A diet high in .... has been linked to high levels of cholesterol in the blood.

c) Explain what can happen if you have too much cholesterol in your body.

....

....

Q4 Gareth did an experiment to compare the elasticity of **arteries** and **veins**. He dissected out an artery and a vein from a piece of fresh meat. He then took a 5 cm length of each vessel, hung different weights on it, and measured how much it stretched. His results are shown in the table.

weight added (g)	length of blood vessel (mm)	
	artery	vein
0	50	50
5	51	53
10	53	56
15	55	59
20	56	-

a) Suggest one way in which he could have decided which was the artery and which was the vein.

....

....

b) Which vessel stretched more easily?

....

c) If Gareth plots his results on a graph, which variable should he put on the vertical (y) axis, and why?

....

d) Why did he take both vessels from the same piece of meat?

....

The Heart

Q1 The diagram below shows the human **heart**, as seen from the front.
The left atrium has been labelled. Complete the remaining labels a) to j).

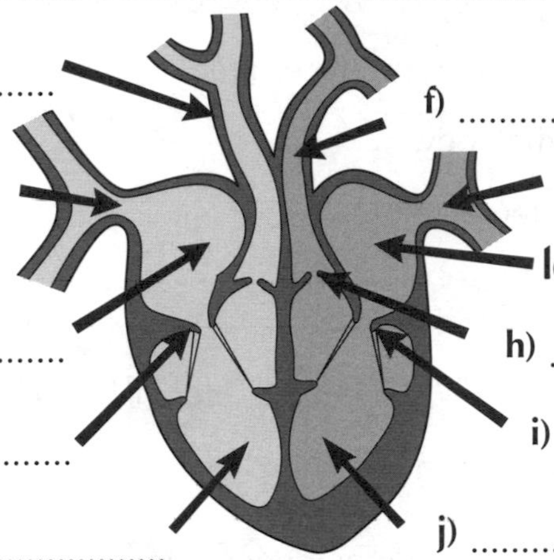

a) ....

b) ....

c) ....

d) ....

e) ....

f) ....

g) ....

left atrium

h) ....

i) ....

j) ....

Q2 Tick the boxes to say whether each statement below is **true** or **false**.

		True	False
a)	Arteries always carry oxygenated blood.	☐	☐
b)	Blood vessels taking blood to and from the lungs are called pulmonary vessels.	☐	☐
c)	The right side of the heart pumps deoxygenated blood.	☐	☐
d)	Valves prevent blood flowing backwards.	☐	☐

Q3 Mammals have a **double** circulatory system in which blood is pumped by the **heart**.

a) Explain the meaning of the term **double circulatory system**.

....

b) Explain why the muscle walls of the atria are thinner than the walls of the ventricles.

....

....

Q4 In extreme cases of heart disease patients may have to undergo a **heart transplant**.
What can doctors do to help prevent the donor heart from being **rejected**?

....

Q5 Complete the passage using the words provided. Each may be used more than once or not at all.

artificial	chambers	vena cava	irregular	pacemaker	valves	regular

The rate at which the heart beats is determined by the .... Sometimes, this stops working properly, and the heartbeat becomes .... In this case, an artificial .... is fitted. Defective heart .... can also be replaced.

The Kidneys and Homeostasis

Q1 Tick the correct boxes to show whether these sentences are **true** or **false**.

		True	False
a)	The kidneys make urea.	☐	☐
b)	Breaking down excess amino acids produces urea.	☐	☐
c)	Proteins can't be stored in the body.	☐	☐
d)	The kidneys monitor blood temperature.	☐	☐
e)	The bladder stores urine.	☐	☐

Q2 One of the kidneys' roles is to adjust the **ion content** of the **blood**.

a) Where do the ions in the blood come from?

....

b) What would happen if the ion content of the blood wasn't controlled?

....

c) Excess ions are removed from the blood by the kidneys. How else can ions be lost from the body?

....

Q3 The kidneys are involved in the control of the body's **water levels**.

a) Name **three** ways that water is lost from the body.

....

b) Complete the table showing how your body maintains a water balance on hot and cold days.

	Do you sweat **a lot** or **a little**?	Is the amount of urine you produce **high** or **low**?	Is the urine you produce **more** or **less** concentrated?
Hot Day			
Cold Day			

c) Sheona ran 25 km. Afterwards she didn't urinate for six hours. When Sheona did urinate, her urine was a very dark colour. Explain why this happened.

....

....

....

Top Tips: You can live with only one kidney and so it is sometimes possible for some people with kidney failure to receive a donated kidney from a living member of their family.

The Pancreas and Diabetes

Q1 Tick the correct boxes to show whether these sentences are **true** or **false**.

		True	False
a)	Insulin must be injected — it can't be taken as a tablet.	☐	☐
b)	If someone with diabetes uses insulin they don't have to be careful what they eat.	☐	☐
c)	The livers of people with diabetes have stopped making insulin.	☐	☐
d)	Injecting insulin regularly will eventually cure a person of diabetes.	☐	☐

Q2 During the 19th century **Banting** and **Best** researched **diabetes** by experimenting on dogs. In some of their experiments they injected an extract into diabetic dogs.

a) Where did they get the extract from? ....

b) When they injected the extract into a diabetic dog, its blood sugar level changed.

i) Describe how the blood sugar level changed.

....

ii) Which hormone did the extract contain? ....

Q3 Describe the **improvements** that have been made in the treatment of diabetes in the areas below:

a) The source of the insulin used by diabetics.

....

....

b) The way diabetics take their insulin.

....

....

Q4 Injecting insulin can be **painful** and **inconvenient**.

a) What **surgical** treatment can be used to cure type 1 diabetes? ....

b) Describe some of the problems with the treatment you have named.

....

....

c) Scientists are constantly researching new treatments and cures for diabetes.
Name **two** treatments that are currently in development.

1. ....

2. ....

Top Tips: Many diabetics are able to control their blood sugar levels and lead normal lives. Sir Steve Redgrave won a gold medal at the Olympics after being diagnosed with diabetes.

Growth

Q1 Daniel wants to measure the **growth** of his new puppy.

a) Suggest two measurements he could take to record the puppy's **size**.

....

b) He also records the puppy's wet weight. What is wet weight?

....

c) Give a disadvantage of measuring growth by recording an organism's wet weight.

....

Q2 Describe two examples of animals that are able to grow **new limbs**.

1. ....

2. ....

Q3 Give two differences in **growth** between plants and animals.

1. ....

2. ....

Q4 Zapphites have a similar growth pattern to humans. The graph shows the **head circumference** of baby Zapphites between birth and 40 weeks. The shaded area shows where 90% of babies fall.

Baby's name	Age (weeks)	Head circumference (cm)
Charles	19	46.5
Edward	15	51
Engletree	34	54
George	27	49
Henry	23	49
Oliver	29	57
Richard	39	50
Xionbert	23	51.5

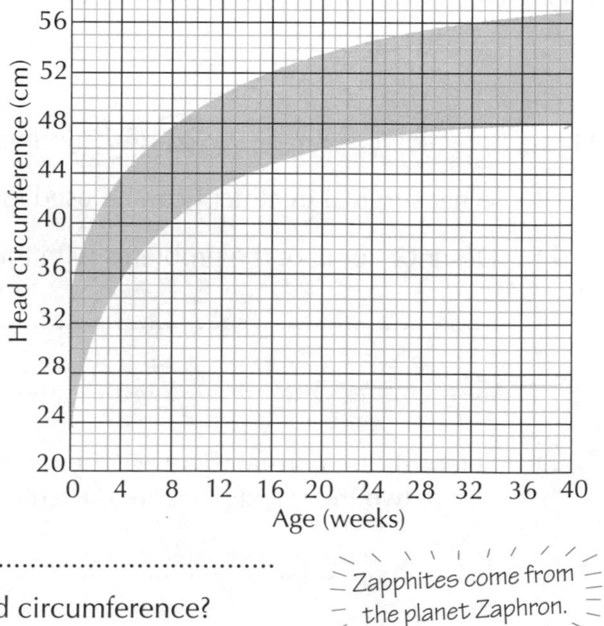

a) Plot the head circumferences of the above babies.

b) Which baby's size may cause concern? ....

c) Why might Zapphite doctors monitor a baby's head circumference?

Zapphites come from the planet Zaphron.

....

....

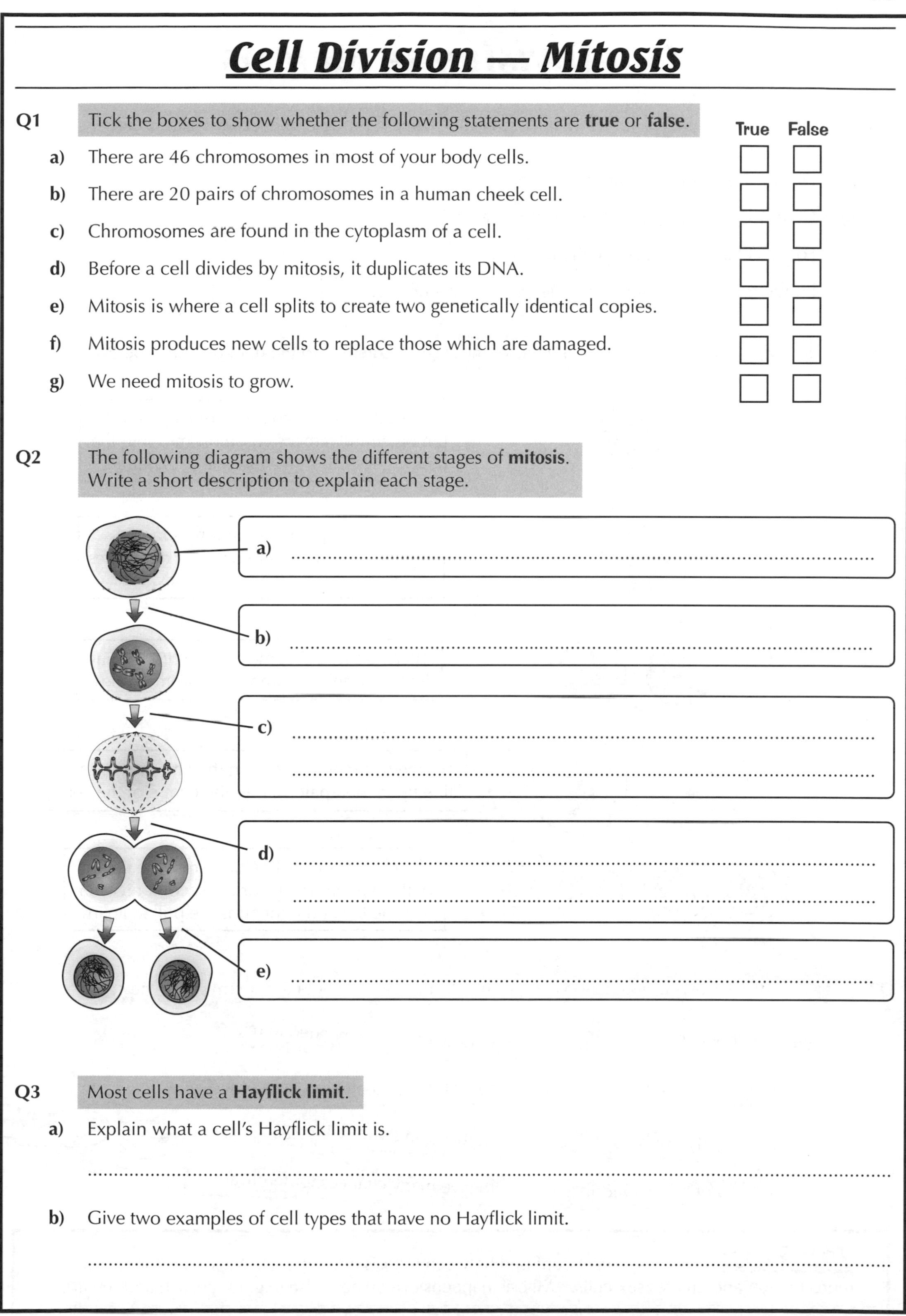

Cell Division — Mitosis

Q1 Tick the boxes to show whether the following statements are **true** or **false**.

		True	False
a)	There are 46 chromosomes in most of your body cells.	☐	☐
b)	There are 20 pairs of chromosomes in a human cheek cell.	☐	☐
c)	Chromosomes are found in the cytoplasm of a cell.	☐	☐
d)	Before a cell divides by mitosis, it duplicates its DNA.	☐	☐
e)	Mitosis is where a cell splits to create two genetically identical copies.	☐	☐
f)	Mitosis produces new cells to replace those which are damaged.	☐	☐
g)	We need mitosis to grow.	☐	☐

Q2 The following diagram shows the different stages of **mitosis**.
Write a short description to explain each stage.

a) ....

b) ....

c) ....

....

d) ....

....

e) ....

Q3 Most cells have a **Hayflick limit**.

a) Explain what a cell's Hayflick limit is.

....

b) Give two examples of cell types that have no Hayflick limit.

....

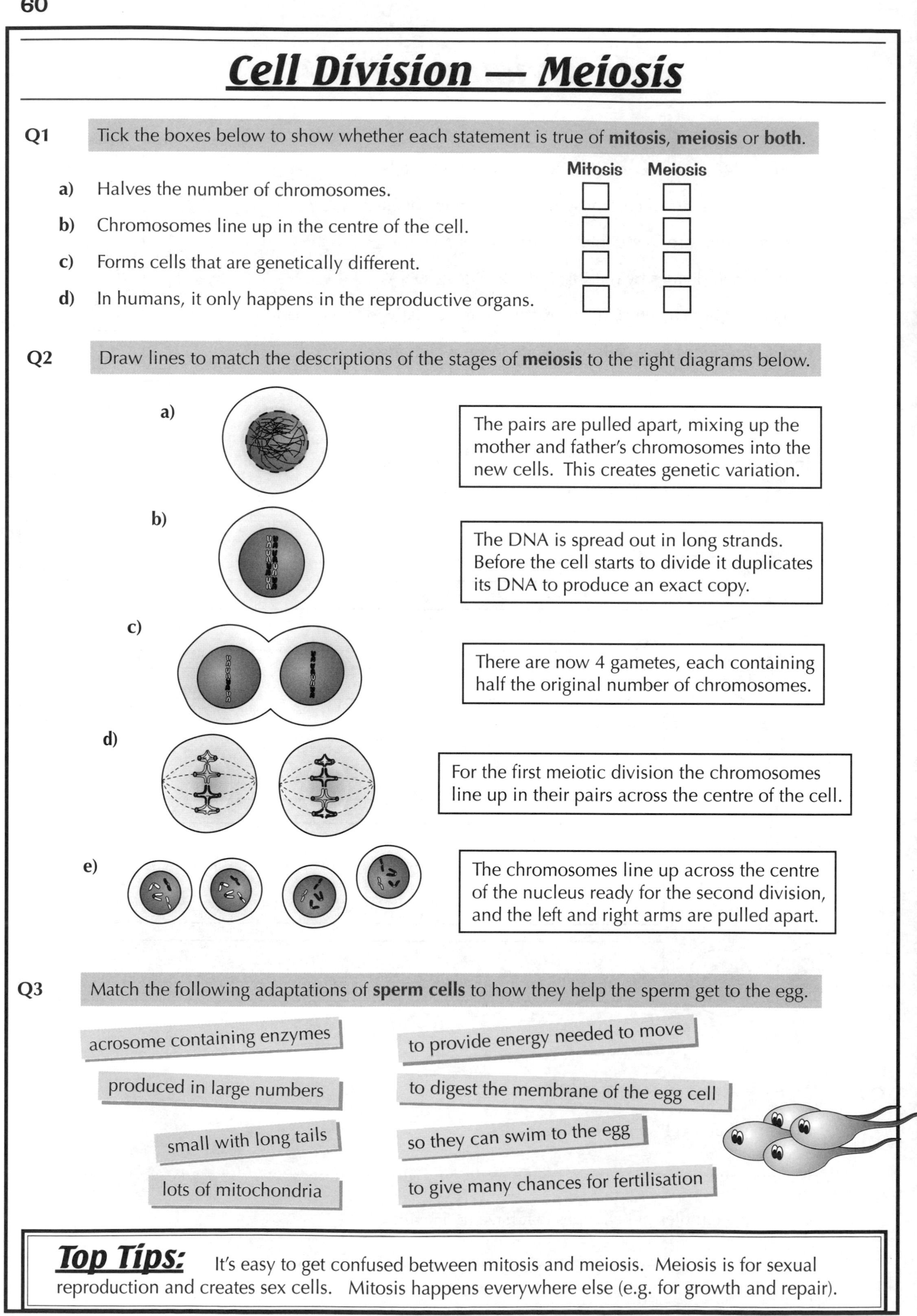

Cell Division — Meiosis

Q1 Tick the boxes below to show whether each statement is true of **mitosis**, **meiosis** or **both**.

		Mitosis	Meiosis
a)	Halves the number of chromosomes.	☐	☐
b)	Chromosomes line up in the centre of the cell.	☐	☐
c)	Forms cells that are genetically different.	☐	☐
d)	In humans, it only happens in the reproductive organs.	☐	☐

Q2 Draw lines to match the descriptions of the stages of **meiosis** to the right diagrams below.

a) b) c) d) e)

The pairs are pulled apart, mixing up the mother and father's chromosomes into the new cells. This creates genetic variation.

The DNA is spread out in long strands. Before the cell starts to divide it duplicates its DNA to produce an exact copy.

There are now 4 gametes, each containing half the original number of chromosomes.

For the first meiotic division the chromosomes line up in their pairs across the centre of the cell.

The chromosomes line up across the centre of the nucleus ready for the second division, and the left and right arms are pulled apart.

Q3 Match the following adaptations of **sperm cells** to how they help the sperm get to the egg.

acrosome containing enzymes

produced in large numbers

small with long tails

lots of mitochondria

to provide energy needed to move

to digest the membrane of the egg cell

so they can swim to the egg

to give many chances for fertilisation

Top Tips: It's easy to get confused between mitosis and meiosis. Meiosis is for sexual reproduction and creates sex cells. Mitosis happens everywhere else (e.g. for growth and repair).

Sexual Reproduction — Ethics

Q1 Human pregnancies may be **terminated**.

a) What is the legal limit for a termination?

....

b) Explain why the limit was set at this stage of the pregnancy in Britain.

....

....

c) Give two situations when a termination may be carried out later than this date.

1. ....

2. ....

d) Give an argument against abortion at any stage of a pregnancy.

....

....

e) Why do some people feel that the legal limit for abortion should be changed?

....

....

Q2 During **in vitro fertilisation** (IVF) a cell can be removed from an embryo and **screened** for genetic disorders like Huntington's disease. If a faulty allele is present, the embryo is destroyed.

a) Explain why some people think embryo screening is a **bad** thing.

....

....

....

b) Explain why some people think embryo screening is a **good** thing.

....

....

....

Top Tips: A lot of people feel very strongly about these issues. It's important for you to understand all the different views and arguments, not only the ones you agree with.

Stem Cells and Differentiation

Q1 The following terms are related to **stem cells**. Explain what each term means.

a) specialised cells ....

b) differentiation ....

c) undifferentiated cells ....

....

Q2 How are **embryonic** stem cells different from **adult** stem cells?

....

....

....

Q3 Describe a way that stem cells are **already** used in **medicine**.

....

....

....

Q4 In the future, **embryonic stem cells** might be used to replace faulty cells in sick people. Match the problems below to the potential cures which could be made with stem cells.

diabetes	heart muscle cells
paralysis	insulin-producing cells
heart disease	brain cells

Q5 People have **different opinions** when it comes to **stem cell research** using embryos.

a) Give one argument **in favour** of embryonic stem cell research.

....

....

b) Give one argument **against** embryonic stem cell research.

....

....

Growth in Plants

Q1 Tick the boxes to show whether the following statements are **true** or **false**.

	True	False
a) Plant shoots grow away from light.	☐	☐
b) Plant roots grow towards water.	☐	☐
c) Plant roots grow away from gravity.	☐	☐
d) If the tip of a shoot is removed, the shoot may stop growing.	☐	☐

Q2 Cedrick placed some **seedlings** on the surface of damp soil which was exposed to **light** from a lamp. The appearance of the seedlings is shown in the diagram.

start: bean, shoot, root

5 days later: bean, shoot, root

a) What **hormones** are responsible for these changes?

....

b) Where are these hormones produced?

....

c) Explain the results observed with the:

i) shoot. ....

....

ii) root. ....

....

Q3 Ronald owns a fruit farm which grows **seedless satsumas**. The fruit is picked **before** it is ripe and transported to a market.

fruit picked → fruit packaged → fruit transported to market → fruit displayed

a) Explain how satsumas can be grown without pips.

....

b) Suggest why the satsumas are picked before they are ripe.

....

c) How will the unripened satsumas be ripened in time to reach the market?

....

Selective Breeding

Q1 Garfield wants to breed one type of plant for its **fruit**, and another as an **ornamental house plant**.

Suggest **two** characteristics that would be desirable in each kind of plant.

Fruit plant: ....

Ornamental house plant: ....

Q2 Describe two **disadvantages** of selective breeding.

1. ....

....

2. ....

....

Q3 Describe how **selective breeding** could be used to improve the following:

a) The number of offspring in sheep. ....

....

b) The yield from dwarf wheat. ....

....

Q4 The graph shows the **milk yield** for a population of cows over three generations.

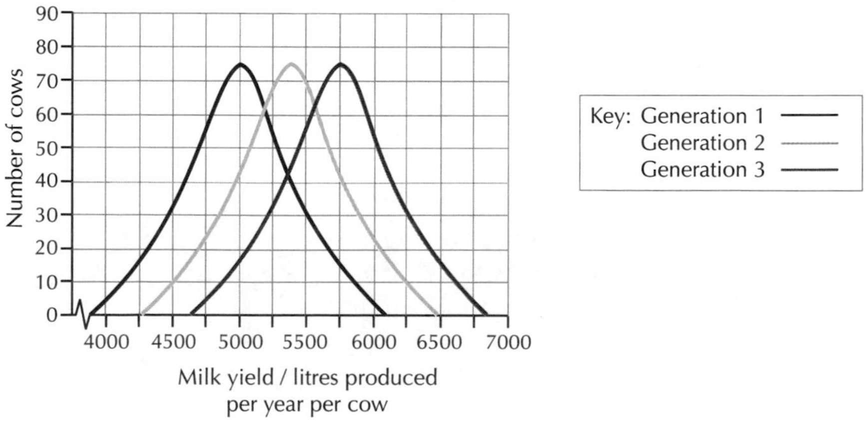

a) Do you think that selective breeding is likely to have been used in these cows? Explain your answer.

....

b) What is the increase in the average milk yield per cow from generation 1 to generation 2?

....

Top Tips: So, if you wanted to take over the world using goldfish, you would probably want to breed together the more aggressive goldfish with long memories, rather than the dappy ones that just idly swim around in a circle all day long. (You can tell which ones are aggressive — they bite.)

Adult Cloning

Q1 **Dolly** the sheep was cloned from an adult cell.

a) Write the correct letter (A, B, C or D) next to each label below to show where it belongs on the diagram.

Egg cell
Adult body cell
A
B
C
D
Sheep

removing and discarding a nucleus ....

implantation in a surrogate mother ....

useful nucleus extracted ....

formation of a diploid cell ....

b) What type of cell division does the egg use to divide? ....

c) Give a risk associated with this type of cloning.

....

Q2 Some animals can be genetically engineered to produce **human blood clotting agents**.

a) Why would it be useful to be able to clone these animals?

....

....

b) Suggest why some people might be reluctant to use medicines produced by cloned animals.

....

Q3 Adult cloning may help to make **xenotransplantation** safe.

a) What is xenotransplantation?

....

b) Explain the role that cloning could play in xenotransplantation.

....

....

Q4 Summarise the **ethical** issues involved in cloning humans.

....

....

....

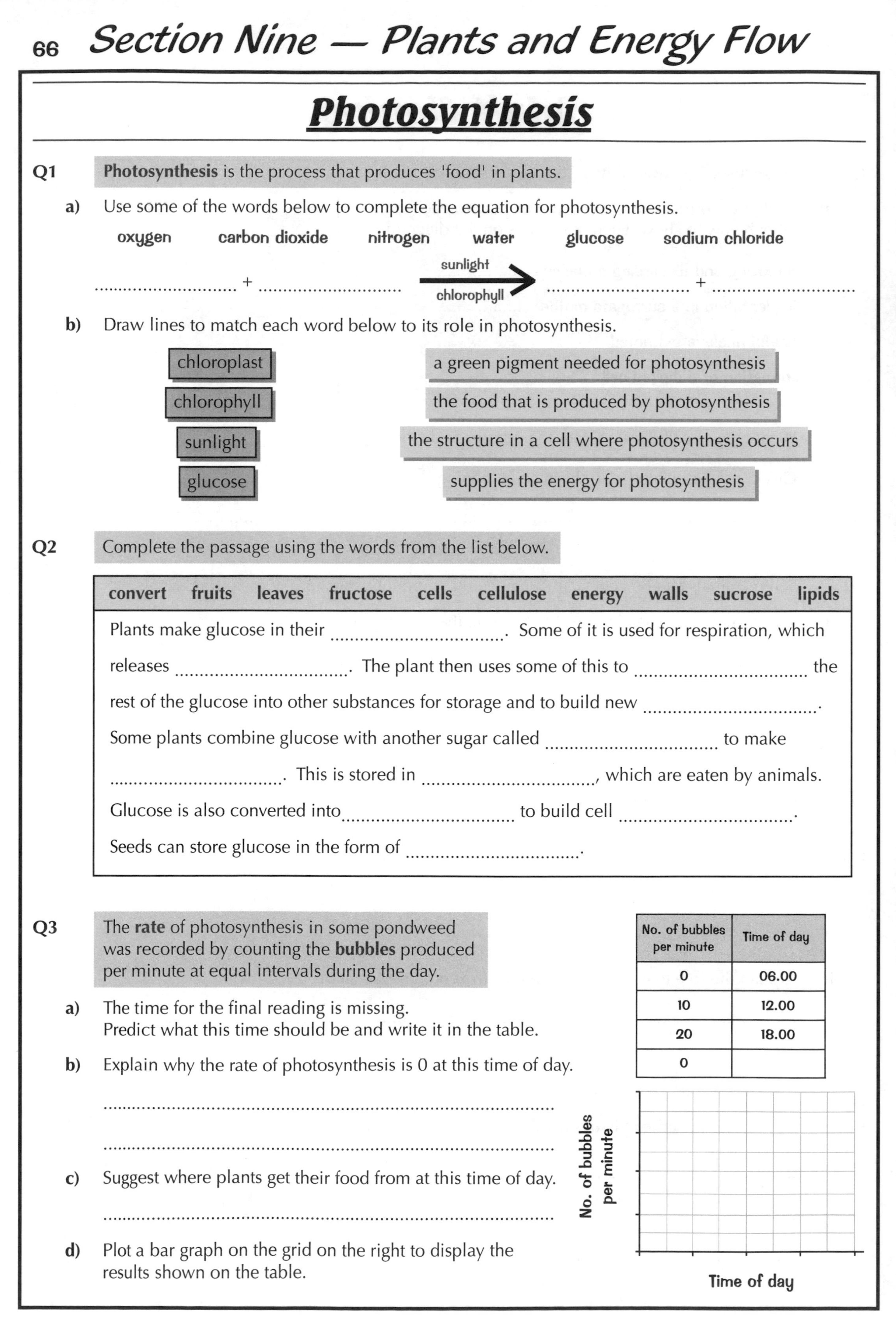

Photosynthesis

Q1 **Photosynthesis** is the process that produces 'food' in plants.

a) Use some of the words below to complete the equation for photosynthesis.

oxygen **carbon dioxide** **nitrogen** **water** **glucose** **sodium chloride**

.... + ... $\xrightarrow[\text{chlorophyll}]{\text{sunlight}}$ .... + ...

b) Draw lines to match each word below to its role in photosynthesis.

chloroplast	a green pigment needed for photosynthesis
chlorophyll	the food that is produced by photosynthesis
sunlight	the structure in a cell where photosynthesis occurs
glucose	supplies the energy for photosynthesis

Q2 Complete the passage using the words from the list below.

convert **fruits** **leaves** **fructose** **cells** **cellulose** **energy** **walls** **sucrose** **lipids**

Plants make glucose in their .... Some of it is used for respiration, which releases .... The plant then uses some of this to .... the rest of the glucose into other substances for storage and to build new ....

Some plants combine glucose with another sugar called .... to make .... This is stored in ...., which are eaten by animals.

Glucose is also converted into................................. to build cell ....

Seeds can store glucose in the form of ....

Q3 The **rate** of photosynthesis in some pondweed was recorded by counting the **bubbles** produced per minute at equal intervals during the day.

No. of bubbles per minute	Time of day
0	06.00
10	12.00
20	18.00
0	

a) The time for the final reading is missing. Predict what this time should be and write it in the table.

b) Explain why the rate of photosynthesis is 0 at this time of day.

....

....

c) Suggest where plants get their food from at this time of day.

....

d) Plot a bar graph on the grid on the right to display the results shown on the table.

Rate of Photosynthesis

Q1 Below are some straightforward questions about **limiting factors**. Hooray.

a) List **three** factors that can limit the rate of photosynthesis.

1. .... 2. .... 3. ....

b) Explain the meaning of the term "limiting factor".

....

c) The limiting factor at a particular time depends on the environmental conditions, e.g. season (such as winter). Name two other environmental conditions that may affect the rate of photosynthesis.

1. .... 2. ....

Q2 The table shows the average daytime summer **temperatures** in different habitats around the world.

Habitat	Temperature (°C)
Forest	19
Arctic	0
Desert	32
Grassland	22
Rainforest	27

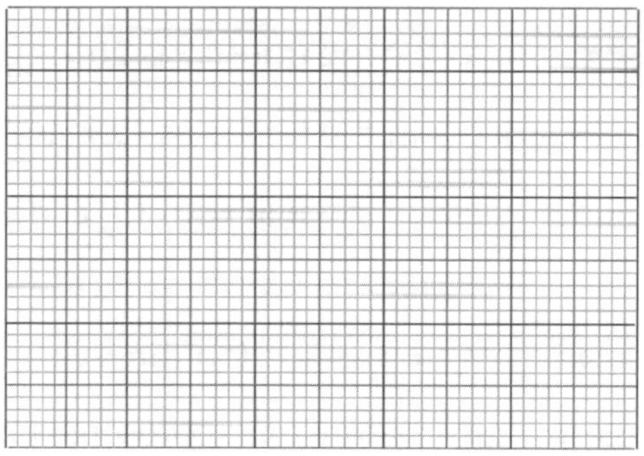

a) Plot a **bar chart** for these results on the grid.

b) From the values for temperature, in which area would you expect fewest plants to grow?

....

c) Suggest a reason for your answer above using the terms **enzymes** and **photosynthesis**.

....

....

Q3 Seth investigated the effect of different concentrations of **carbon dioxide** on the rate of photosynthesis of his Swiss cheese plant. He measured the rate of photosynthesis with increasing light intensity at **three** different CO_2 concentrations. The results are shown on the graph below.

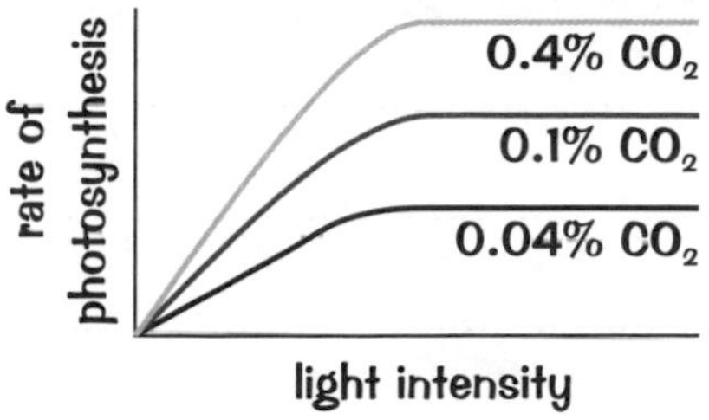

rate of photosynthesis

0.4% CO_2

0.1% CO_2

0.04% CO_2

light intensity

a) What effect does increasing the concentration of CO_2 have on the rate of photosynthesis?

....

....

....

b) Explain why all the graphs level off eventually.

....

....

Think about the third limiting factor.

Leaf Structure

Q1 Name the parts labelled **A – E** to complete the diagram of a **leaf** below.

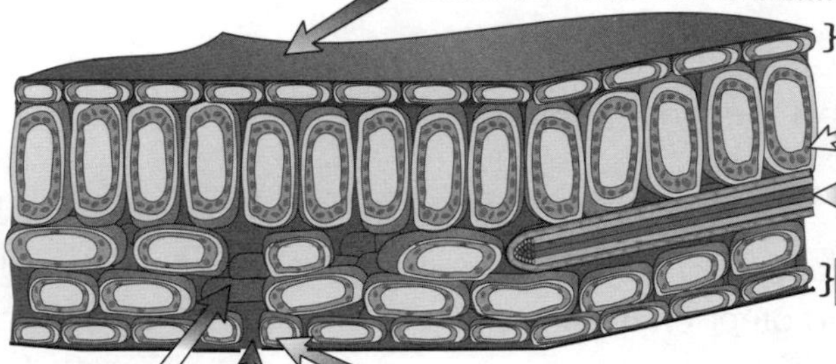

C ....

} B ....

A ....

....

spongy mesophyll layer

chloroplast

vein

} lower epidermis

air space

E ....

D ....

Q2 Answer the following questions about **gas exchange** in leaves.

a) Which process in the leaf uses CO_2 and produces O_2? ....

b) Which process in the leaf uses O_2 and produces CO_2? ....

Q3 A diagram of a cross-section through part of a **leaf** which is **photosynthesising** is shown.

a) Suggest what substance is represented by each of the letters shown on the diagram.

A ....

B ....

C ....

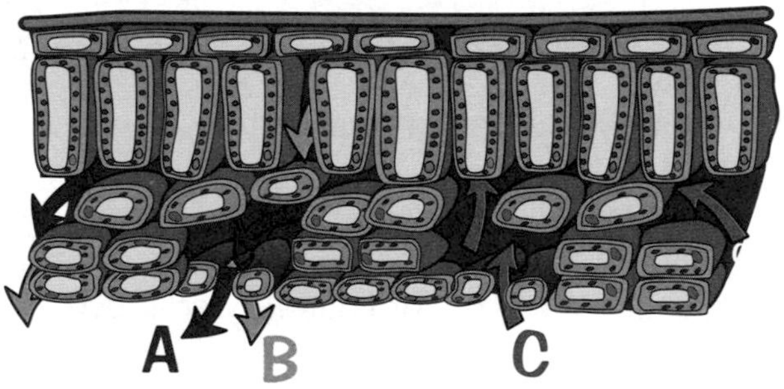

A B C

b) What differences in gas exchange would there be at night? Explain your answer.

....

....

Q4 Describe how the following features of leaves help with **photosynthesis**.

a) Air spaces in the mesophyll layer. ....

b) Broad leaves. ....

c) Veins. ....

d) Lots of chloroplasts. ....

Top Tips: Leaves are adapted for efficient diffusion. They increase gas exchange by creating a large surface area and a short distance for diffusion to happen across. They have also adapted to minimise water loss from the plants. The clever little things...

Transpiration

Q1 Complete this diagram of a **plant** according to the instructions given below.

a) Draw an **X** on the diagram to show where water enters the plant.

b) Add **Y**s to the diagram to show where water leaves the plant.

c) Add arrows to the diagram to show how water moves from where it enters to where it leaves.

Q2 Tick the boxes to show whether each of the following statements is **true** or **false**.

		True	False
a)	The transpiration rate decreases as the temperature increases.	☐	☐
b)	The more intense the light, the faster the transpiration rate.	☐	☐
c)	Transpiration happens more slowly when the air is humid.	☐	☐
d)	As the wind speed increases, the rate of transpiration decreases.	☐	☐

Q3 Use words from the box to complete the passage below. Each word may be used more than once.

osmosis leaves evaporation roots flowers phloem diffusion transpiration xylem stem

Water leaves plants through the .... by the processes of .... and .... This creates a shortage of water in the ...., which draws water from the rest of the plant through the .... vessels. This causes more water to be drawn up from the .... The whole process is called ....

Q4 Give three ways that transpiration **benefits** plants.

....

....

Q5 Stomata are different sizes at different **light intensities**.

a) Would you expect a plant's stomata to be open or closed on a sunny morning? Explain why.

....

b) What happens to the stomata at night? What is the advantage of this?

....

c) If the supply of water to the roots of a plant dries up, the stomata close.
Give one **advantage** and one **disadvantage** of this mechanism for the plant.

....

....

Water Flow in Plants

Q1 Plants need to **balance** water loss with water gain.

a) Give two ways that plants are adapted to reduce water loss from their leaves.

1. ....

2. ....

b) How have plants in hot climates adapted to reduce water loss?

....

Q2 Plant cells look different depending on how much **water** they contain.

a) Use the words in the box on the right to describe the states of the following cells.

plasmolysed	turgid	normal	flaccid

A .... B .... C .... D ....

b) Explain why plants start to wilt if they don't have enough water.

....

....

c) Explain why the cell in diagram D hasn't totally lost its shape.

....

Q3 **Xylem** is designed for transporting substances in plants.

a) Name **two** things that are transported by the xylem. ....

b) Give another function of the xylem, other than transport. ....

c) How is the xylem adapted for this other function?

....

Q4 Put the following statements under the **correct heading** in the table.

- transport water
- made of living cells
- have end-plates
- have no end-plates
- transport food
- made of dead cells

XYLEM VESSELS	PHLOEM VESSELS

Minerals for Healthy Growth

Q1 Draw lines to match the following **minerals** with their **functions** in plants.

MAGNESIUM	for making proteins
POTASSIUM	for making chlorophyll
PHOSPHATES	for making DNA and cell membranes
NITRATES	for helping enzymes to function

Q2 A diagram of a **specialised plant cell** is shown.

a) Name the type of cell shown. ....

b) What is the main function of this type of cell?

....

c) How is this type of cell adapted for its function?

....

d) Explain why minerals are **not** absorbed from the soil by **diffusion**.

....

....

e) Explain how these specialised cells absorb minerals from the soil.

....

Q3 Spring has arrived but Pat has noticed that his **grain crop** has **stunted growth** and **yellow older leaves**. He has grown grain on this field for the last **three years**.

a) Which mineral would you recommend that Pat add to ensure better growth of his crops?

....

Pat has been offered some **manure** for his field. The table shows the mineral content of different manures.

Material	% Nitrogen	% Phosphorus	% Potassium
Bullock manure	0.6	0.1	0.7
Cow manure	0.4	0.1	0.4
Horse manure	0.6	0.1	0.5
Pig manure	0.4	0.1	0.5
Poultry manure	1	0.4	0.6
Sheep manure	0.8	0.1	0.7

b) Which type of manure would you recommend that Pat use? Explain your answer.

....

....

Top Tips: Whatever you do, don't say in an exam that minerals enter the root by diffusion. That would be impossible, because there is a low level of minerals in the soil but a lot inside the root. Active transport has to use energy to drag those minerals kicking and screaming in the wrong direction.

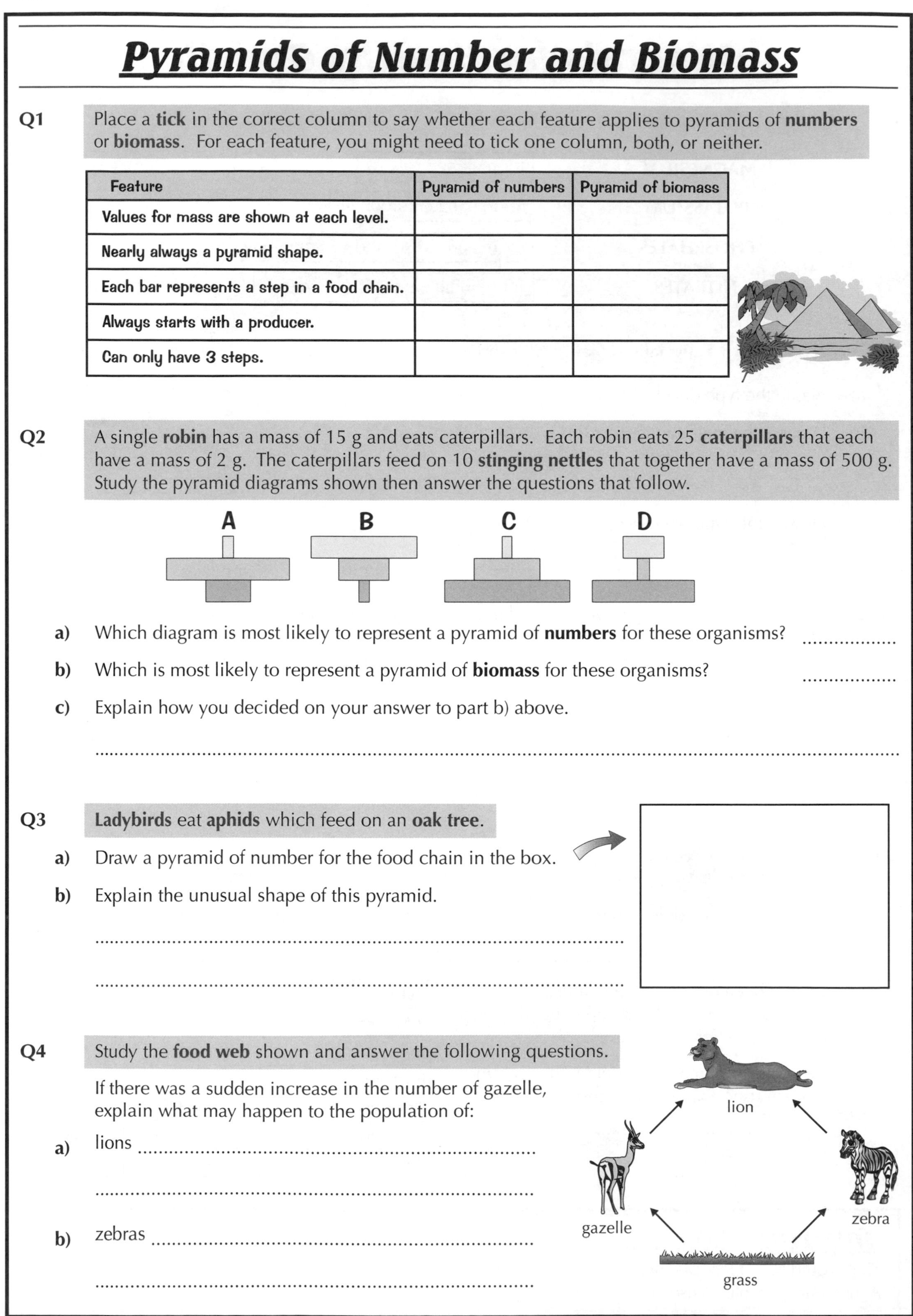

Pyramids of Number and Biomass

Q1 Place a **tick** in the correct column to say whether each feature applies to pyramids of **numbers** or **biomass**. For each feature, you might need to tick one column, both, or neither.

Feature	Pyramid of numbers	Pyramid of biomass
Values for mass are shown at each level.		
Nearly always a pyramid shape.		
Each bar represents a step in a food chain.		
Always starts with a producer.		
Can only have 3 steps.		

Q2 A single **robin** has a mass of 15 g and eats caterpillars. Each robin eats 25 **caterpillars** that each have a mass of 2 g. The caterpillars feed on 10 **stinging nettles** that together have a mass of 500 g. Study the pyramid diagrams shown then answer the questions that follow.

a) Which diagram is most likely to represent a pyramid of **numbers** for these organisms? ....

b) Which is most likely to represent a pyramid of **biomass** for these organisms? ....

c) Explain how you decided on your answer to part b) above.

....

Q3 **Ladybirds** eat **aphids** which feed on an **oak tree**.

a) Draw a pyramid of number for the food chain in the box.

b) Explain the unusual shape of this pyramid.

....

....

Q4 Study the **food web** shown and answer the following questions.

If there was a sudden increase in the number of gazelle, explain what may happen to the population of:

a) lions ....

....

b) zebras ....

....

Energy Transfer and Energy Flow

Q1 Complete the sentences below by circling the correct words in each pair.

a) Nearly all life on Earth depends on **food** / **energy** from the Sun.

b) **Plants** / **Animals** can make their own food by a process called **photosynthesis** / **respiration**.

c) To obtain energy animals must **decay** / **eat** plant material or other animals.

d) Animals and plants release energy through the process of **photosynthesis** / **respiration**.

e) Some of the energy released in animals is **gained** / **lost** through **growth** / **movement** before it reaches organisms at later steps of the food chain.

f) Some energy is lost between steps of a food chain because it's used to make **edible** / **inedible** materials such as **hair** / **flesh**.

Q2 An **aquatic food chain** is shown.

plankton → shrimp → small fish → carp

100 000 kJ .... 1000 kJ ....

a) 90 000 kJ is lost at the first transfer.

i) Write the amount of energy available in the shrimps for the small fish in the space provided.

ii) Calculate the **efficiency** of energy transfer from the plankton to the **shrimps**.

....

b) The energy transfer from the small fish to the carp is **5%** efficient.

i) Write the amount of **energy** passed on to the **carp** in the space provided above.

ii) How much energy is **lost** from the food chain at this stage?

....

Q3 Study the diagram of **energy transfer** shown.

a) Using the figures shown on the diagram, work out the percentage of the Sun's energy that is passed to the grass.

....

b) Only 10% of the energy in the grass reaches the next trophic level. Work out how much energy from the grass passes to animal A.

....

c) **B** and **C** are processes that represent energy loss. Suggest what these processes might be.

....

d) Why do food chains rarely have more than five trophic levels?

....

....

Biomass and Fermentation

Q1 Give four ways that **energy** stored as **biomass** can be released for human use.

1. .... 2. ....

3. .... 4. ....

Q2 Answer the following questions about **biofuels**.

a) Explain why burning and replanting trees doesn't contribute to global warming.

....

b) What piece of equipment is used to make biogas?

....

c) How is biogas used?

....

Q3 Tick the boxes to show whether each of the following statements is **true** or **false**.

	Statement	True	False
a)	The culture medium is a solid.	☐	☐
b)	The food for the microorganisms is contained in the air.	☐	☐
c)	The pH inside the fermenter must be carefully monitored.	☐	☐
d)	It doesn't matter what temperature it is inside the fermenter.	☐	☐
e)	Air is piped in to supply carbon dioxide to the microorganisms.	☐	☐

Q4 The diagram below shows a **fermenter** that can be used for producing **mycoprotein**.

Food in
Microorganisms in
Exhaust gases out
Water out
Water jacket
Paddles
Water in
Product out
Air in

a) What is mycoprotein?

....

b) Give two advantages of using microorganisms to make food.

....

....

c) Explain the purpose of each of the following features of the fermenter:

i) the water jacket ....

ii) the air supply ....

iii) the paddles ....

Top Tips: Not all microorganisms used in fermenters need oxygen — it depends on whether they respire aerobically or anaerobically to produce the useful product.

Managing Food Production

Q1 Three different **food chains** are shown here.

Circle the food chain that shows the most **efficient** production of **food** for **humans** and explain your choice.

Grass → Cattle → Human

Pondweed → Small fish → Salmon → Human

Wheat → Human

....

....

Q2 Give a **disadvantage** of the following methods of improving the efficiency of food production.

a) Animals are crowded together. ....

....

b) Animals are given antibiotics. ....

....

c) Animals are kept warm. ....

....

Q3 Emma compared two ecosystems. **Ecosystem A** was carefully controlled — the fish were kept in large cages and fed a special diet. Pesticides were used to kill unwanted pests. **Ecosystem B** was kept as natural as possible, with no cages, special diet or pest control. Emma's observations are recorded in the table shown.

Time (months)	Number of fish		Average size of fish (mm)		Comments	
	A	B	A	B	A	B
0	200	200	362	348	200 fish introduced.	200 fish introduced.
2	189	191	368	392	A few initial losses due to change in habitat.	A few initial losses due to change in habitat. Initial growth rate seems fast.
4	188	152	374	423	Numbers have stabilised. Water quality good.	High numbers of fish lice. Adults still growing well.
6	277	136	436	426	Breeding looks successful. Fish growth increasing.	Fish lice levels still high. Breeding has started. Growth rate decreasing.
8	349	172	359	372	End of breeding season. Adult fish growing well.	Breeding season. Fish numbers stabilising. Water pH 8.
10	338	184	401	382	Very few of the new fish have been lost.	Breeding season now over. Growth has slowed.
12	336	179	443	393	Population seems stable. Large, healthy fish.	Population stabilising. Water pH improved at 7.5.

a) Suggest why the **average size** of fish drops so much in both ecosystems at 8 months.

....

b) What factors may have affected the **growth rate** and **number** of fish in Ecosystem B?

....

c) What conclusions could Emma draw from her investigation?

....

....

Pesticides and Biological Control

Q1 a) What are **pesticides** and why are they used?

....

....

b) Give one problem that can be caused by the use of pesticides.

....

Q2 **Cockroaches** were sprayed with a **pesticide** to control the size of their population.

Explain what effect this could have on the rest of the food web shown.

cockroach → frog → fox
rabbit → fox

....

....

....

Q3 **Biological control** is an alternative to using pesticides.

a) What is biological control?

....

b) Give two examples of biological control.

1. ....

2. ....

c) Give an **advantage** and a **disadvantage** of using biological control.

Advantage: ....

Disadvantage: ....

Q4 **Pesticides** that were being sprayed onto fields near to a bird of prey's habitat were found in the birds in **toxic** levels.

Birds of prey only eat other animals and fish.

a) Suggest how pesticide that was sprayed onto crops was found in the birds.

....

b) The amount of pesticide sprayed onto the field was carefully controlled to ensure it was at the lowest concentration that would kill the pests. Suggest why the birds contained such large amounts of the pesticide.

....

....

Alternatives to Intensive Farming

Q1 **Hydroponics** is an **alternative method** of growing plants.

a) What are the plants grown in? ....

b) Give an example of a plant that is grown in this way. ....

c) Are the following features of hydroponics an **advantage** or a **disadvantage** of the technique?

i) Small amount of land required. ....

ii) High cost. ....

iii) No weeding required. ....

iv) Support needed by plants. ....

v) Special soluble nutrients needed. ....

Q2 For each of the substances used in **intensive farming** below suggest an **organic farming alternative** and give one **advantage** of the alternative.

a) **Insecticides:** alternative — ....

Advantage: ....

b) **Herbicides:** alternative — ....

Advantage: ....

Q3 Comment on the **cost**, **labour** needed and **environmental** effects for a farmer considering using **manure and compost** instead of artificial fertilisers, and **weeding** instead of herbicides.

a) **Cost** ....

....

....

b) **Amount of labour** ....

....

....

c) **Effect on the environment** ....

....

....

Recycling Nutrients

Q1 Complete the diagram below as instructed to show part of the **carbon cycle**.

CO_2 in the air

plant **animal**

a) Add an arrow or arrows labelled **P** to represent **photosynthesis**.

b) Add an arrow or arrows labelled **R** to represent **respiration**.

c) Add an arrow or arrows labelled **F** to represent **feeding**.

Q2 The sentences below describe how **elements** are **recycled** in a food chain. Sort them into the correct order by numbering them 1 to 5. The first one has been done for you.

- [ ] **Nutrients in plants are passed to animals by feeding and used in respiration to provide energy.**
- [ ] **Materials are recycled and returned to the soil by decay.**
- [1] **Plants take up minerals from the soil.**
- [ ] **Plants use minerals and the products of photosynthesis to make complex nutrients.**
- [ ] **Plants and animals die.**

Q3 **Carbon** is a very important element that is constantly being recycled.

a) How is carbon removed from the atmosphere?

....

b) How is this carbon passed on through the food chain?

....

c) By what process do **all** living organisms return carbon to the air? ....

Q4 Complete the following table about two types of organism that are important in **decay**.

Type of organism involved in decay	Example	How they help in decay
Detritivores	....	....
....	Bacteria / fungi	....

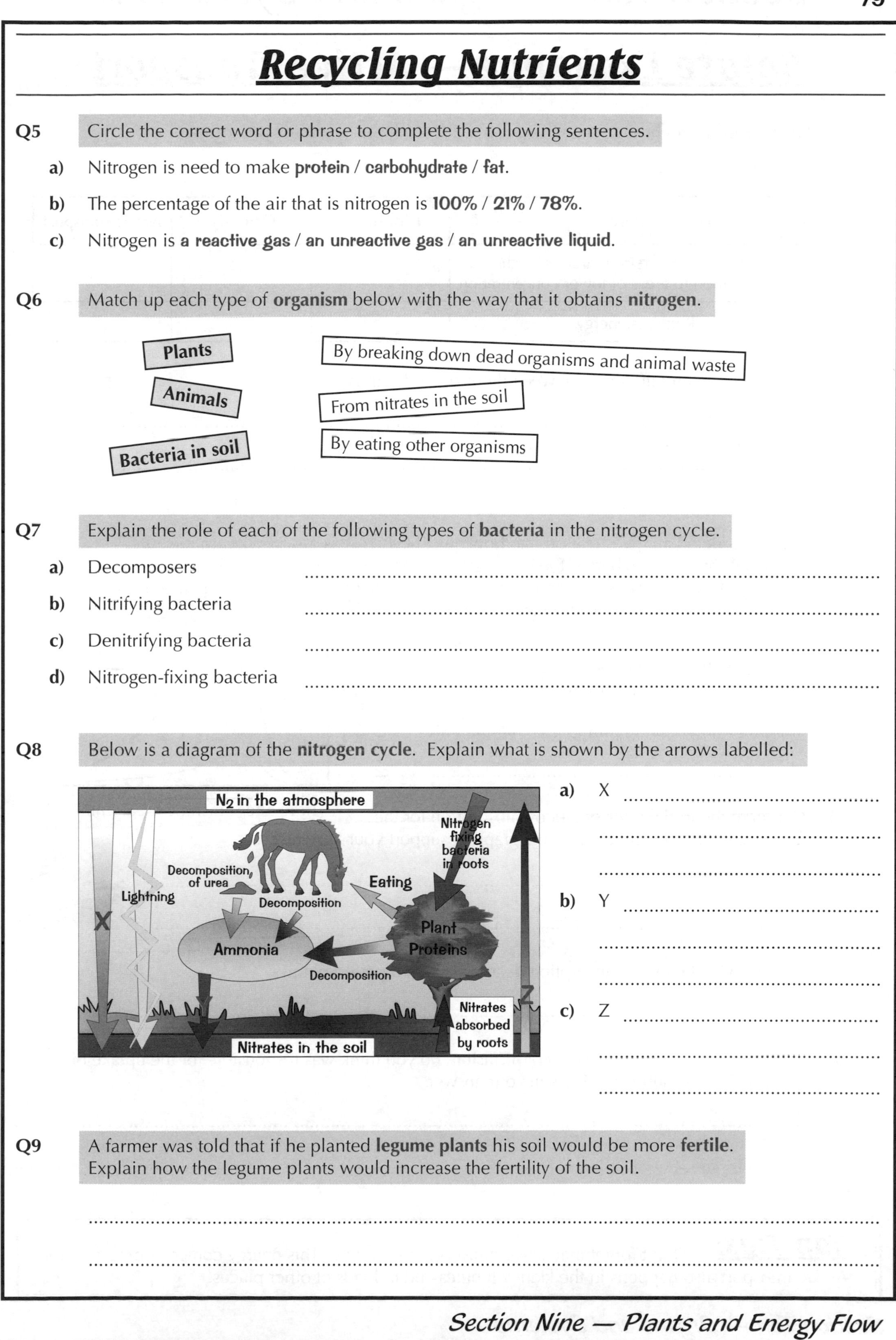

Recycling Nutrients

Q5 Circle the correct word or phrase to complete the following sentences.

a) Nitrogen is need to make **protein** / **carbohydrate** / **fat**.

b) The percentage of the air that is nitrogen is **100%** / **21%** / **78%**.

c) Nitrogen is **a reactive gas** / **an unreactive gas** / **an unreactive liquid**.

Q6 Match up each type of **organism** below with the way that it obtains **nitrogen**.

Plants

Animals

Bacteria in soil

By breaking down dead organisms and animal waste

From nitrates in the soil

By eating other organisms

Q7 Explain the role of each of the following types of **bacteria** in the nitrogen cycle.

a) Decomposers ....

b) Nitrifying bacteria ....

c) Denitrifying bacteria ....

d) Nitrogen-fixing bacteria ....

Q8 Below is a diagram of the **nitrogen cycle**. Explain what is shown by the arrows labelled:

a) X ....

....

....

b) Y ....

....

....

c) Z ....

....

....

Q9 A farmer was told that if he planted **legume plants** his soil would be more **fertile**. Explain how the legume plants would increase the fertility of the soil.

....

....

Solute Exchange — Active Transport

Q1 Substances move through partially permeable membranes by **three** processes.

a) Place ticks in the correct boxes to identify the features of each process.

Feature	Diffusion	Osmosis	Active transport
Substances move from areas of higher concentration to areas of lower concentration			
Requires energy			

b) What is the main difference between diffusion and osmosis?

....

....

Q2 Theo conducted an investigation into how quickly **glucose** is **absorbed** from the gut. He **fasted** for 12 hours before the investigation began and then ate a meal containing a 40 g of **starch** (starch is broken down in the gut into glucose). He then repeated the experiment eating a meal containing 5 g of starch, after fasting.

a) Suggest why there was a 15 minute time delay between eating the meals that contained starch and absorption from the gut.

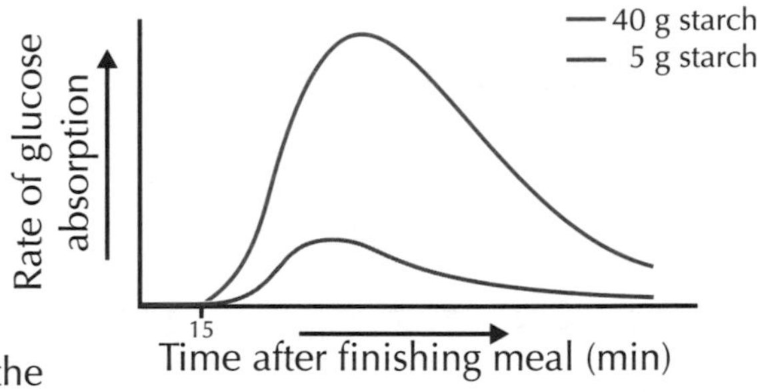

....

....

b) Compare the **initial** rates of glucose absorption for the four meals. Use evidence from the graph to support your answer.

....

....

c) Suggest why the rate of absorption decreased after a period of time.

....

d) Which process, **active transport** or **diffusion**, do you think was responsible for the uptake of glucose in these subjects? Explain your answer.

....

....

Top Tips: Don't forget that active transport uses energy. This energy comes from respiration. Active transport also happens in the kidney tubules and in loads of other places.

The Respiratory System

Q1 a) On the diagram show the positions of the following **structures** by placing the correct letter in the correct box:

A	alveoli	**B**	bronchus
C	trachea	**D**	bronchiole

b) Complete the passages below using the words given. Each word may be used more than once.

out flattens drawn into in diaphragm
up increases decreases down intercostal
forced out of volume relax ribcage

When we breathe .... the .... muscles and the .... contract. This means the diaphragm .... and the ribcage and the sternum move .... and .... This means the volume of the thorax .... in size and the pressure .... Air is then .... the lungs.

Breathing out happens when the intercostal muscles and the diaphragm .... This means that the .... and sternum move .... and .... As a result the .... of the thorax .... and the pressure ...., meaning that air is .... the lungs.

Q2 The respiratory tract is lined with a **mucous membrane**.

a) Complete the diagram by labelling the **goblet cell** and then drawing in the **mucus** and **cilia**.

b) What is the function of the following?

i) Goblet cells ....

ii) Mucus ....

iii) Cilia ....

c) Why is the respiratory tract particularly prone to infection?

....

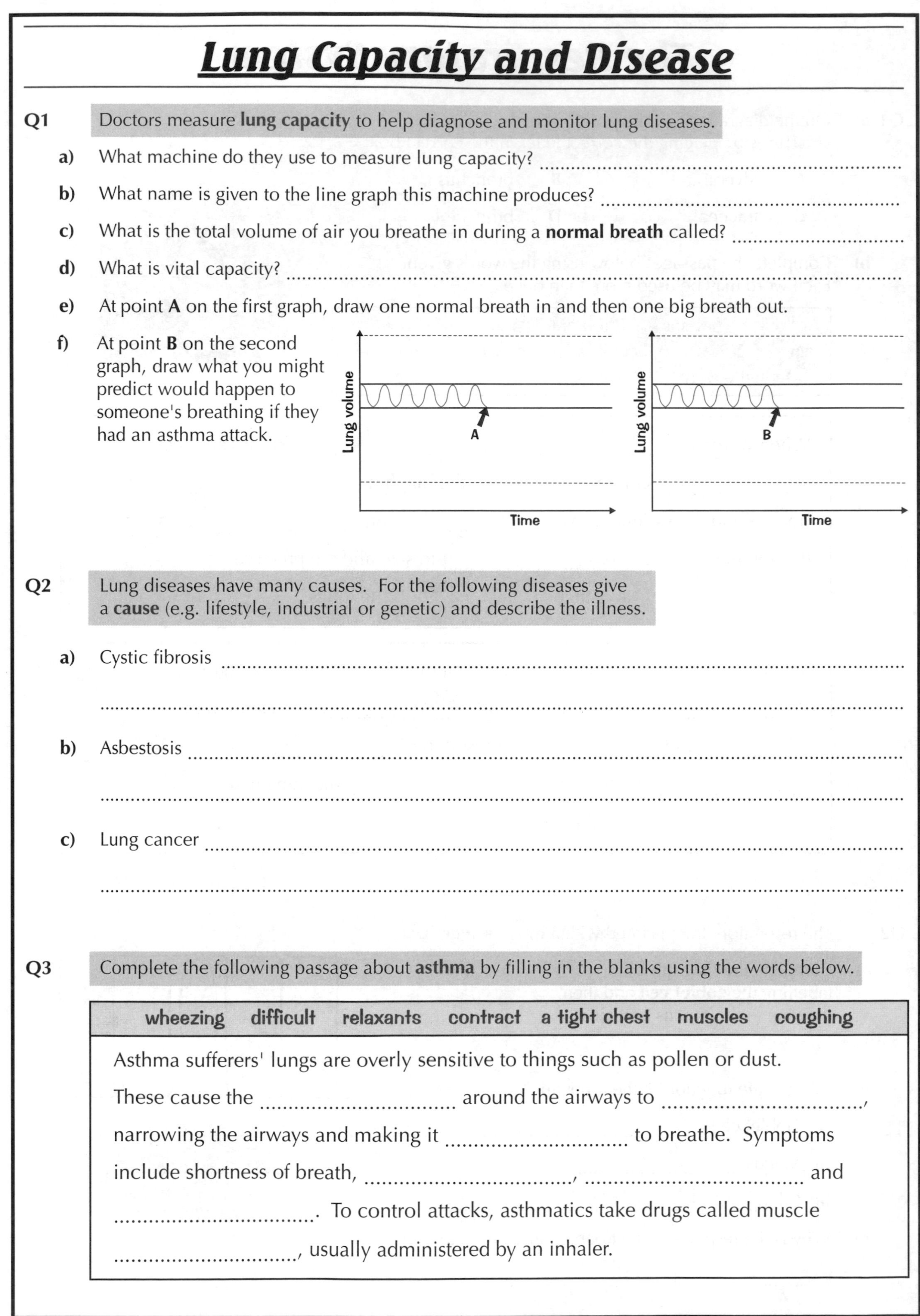

Lung Capacity and Disease

Q1 Doctors measure **lung capacity** to help diagnose and monitor lung diseases.

a) What machine do they use to measure lung capacity? ....

b) What name is given to the line graph this machine produces? ....

c) What is the total volume of air you breathe in during a **normal breath** called? ....

d) What is vital capacity? ....

e) At point **A** on the first graph, draw one normal breath in and then one big breath out.

f) At point **B** on the second graph, draw what you might predict would happen to someone's breathing if they had an asthma attack.

Q2 Lung diseases have many causes. For the following diseases give a **cause** (e.g. lifestyle, industrial or genetic) and describe the illness.

a) Cystic fibrosis ....

....

b) Asbestosis ....

....

c) Lung cancer ....

....

Q3 Complete the following passage about **asthma** by filling in the blanks using the words below.

wheezing difficult relaxants contract a tight chest muscles coughing

Asthma sufferers' lungs are overly sensitive to things such as pollen or dust. These cause the .... around the airways to ...., narrowing the airways and making it .... to breathe. Symptoms include shortness of breath, ..., ... and .... To control attacks, asthmatics take drugs called muscle ...., usually administered by an inhaler.

The Circulation System

Q1 The **heart** is designed to pump **blood** around the body.

a) Explain why the heart is thought of as **two pumps** rather than a single pump.

....

....

b) **i)** Which side of the heart pumps blood to the **lungs**? ....

ii) Is the blood on this side oxygenated or deoxygenated? Circle the correct answer.

oxygenated **deoxygenated**

c) **i)** Which side of the heart pumps blood to the **body**? ....

ii) Is the blood on this side oxygenated or deoxygenated? Circle the correct answer.

oxygenated **deoxygenated**

Think about which side has to pump harder.

d) Suggest why the left side of the heart is more muscular than the right side.

....

....

Q2 The diagram shows the **blood vessels** of the **heart**.

Write the name of each blood vessel beside the letters on the diagram.

A .... B ....

C .... D ....

Right-hand side Left-hand side

Q3 The **cardiac cycle** is the sequence of events that takes place in one complete heartbeat. The sentences below describe some of these events. Number them to show their correct order.

The first one has been done for you.

- [ ] Blood flows into the pulmonary arteries and aorta.
- [1] Blood flows into the atria from the vena cava and pulmonary veins.
- [ ] The ventricles contract.
- [ ] The cycle starts again as blood flows into the atria.
- [ ] The atria contract pushing blood into the ventricles.

The Heart and Heart Disease

Q1 Heart rate is controlled by **pacemaker cells.**

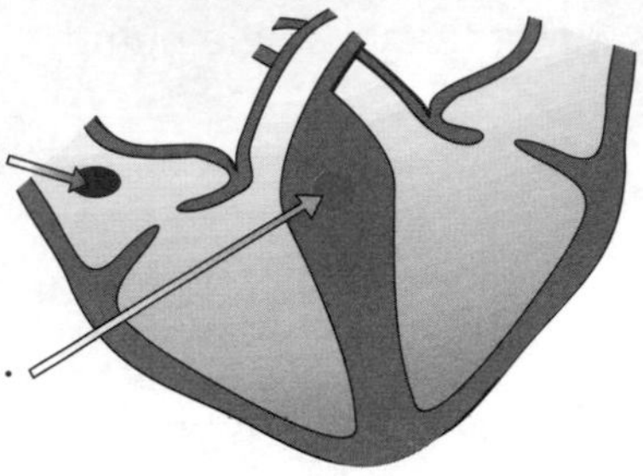

....

....

a) Label the diagram to show where the SAN is.

b) Label the diagram to show where the AVN is.

c) Draw arrows on the diagram to show the direction(s) in which the electrical impulses from the AVN travel.

Q2 The diagrams below show **ECG recordings** from a healthy person and a person admitted to hospital with a heart problem.

a) What do the letters ECG stand for? ....

b) What is recorded by the ECG trace?

....

c) Which parts of the cardiac cycle do the labels **A**, **B**, **C** and **D** represent?

A ....

B ....

C ....

D ....

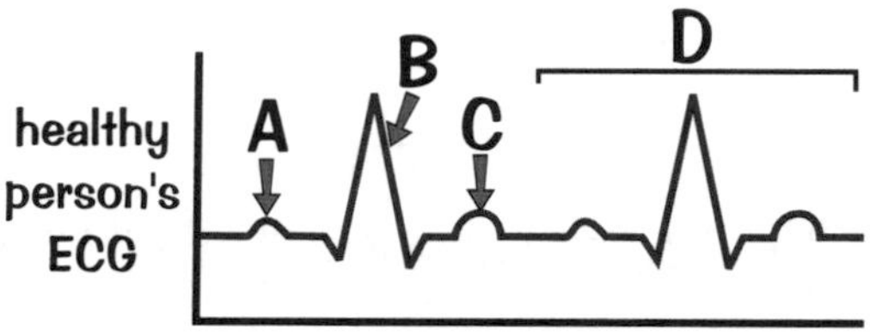

healthy person's ECG

A B C D

d) How does the ECG trace of the hospital patient **differ** from that of the healthy person?

....

....

....

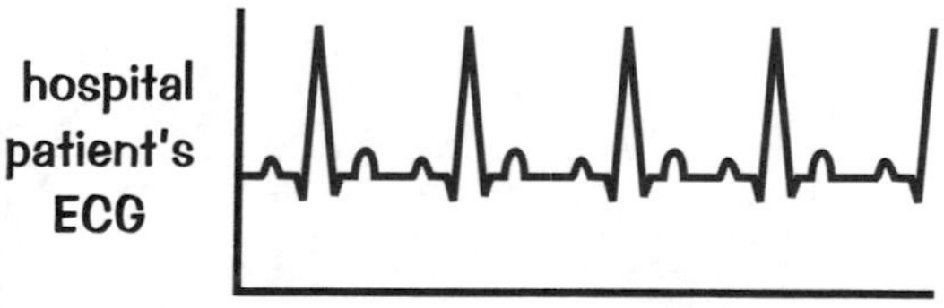

hospital patient's ECG

e) Name another test used to investigate heart function. ....

Q3 There are several common factors that make **heart disease** more likely.

a) List **five** lifestyle 'risk factors' that can increase the likelihood of heart disease developing.

....

....

b) Why is it unhealthy to have too much **saturated fat** in your diet?

....

....

Blood

Q1 If you cut yourself, your blood **clots** so that you don't lose too much. How well blood clots is affected by substances in your **diet**.

a) i) How is a blood clot formed? ....

....

ii) Name a protein that is involved in clotting. ....

b) Which vitamin is needed for blood to clot properly? ....

c) Which type of food contains high levels of the above vitamin? ....

d) What effect does alcohol have on blood clotting?

....

Q2 Use the words below to complete the passage about **blood clotting disorders**.

heparin	haemophilia	strokes	warfarin	clotting factor	DVT	aspirin

.... and .... are conditions caused by excessive blood clotting. People at risk of these conditions take drugs such as ...., ... and .... to help prevent blood clotting. .... is a genetic disease in which the blood takes longer to clot. It is caused by the absence of a .... and is treated by injecting it.

Q3 The presence or absence of blood group antigens determines who can receive **blood transfusions** from certain donors.

a) In what situation might you need a blood transfusion?

....

b) Circle the right answers to show whether a person can receive or donate a particular blood group.

i) If they are blood group **A** can they **receive** blood group **B**? **yes / no**

ii) If they are blood group **A** can they **receive** blood group **O**? **yes / no**

iii) If they are blood group **AB** can they **receive** blood group **B**? **yes / no**

iv) If they are blood group **AB** can they **donate** to blood group **B**? **yes / no**

c) Which blood group can only receive blood of the same blood type? ....

d) Explain why a person with blood group A can't accept blood from a person with blood group B.

....

....

Waste Disposal — The Kidneys

Q1 The diagram shows the steps that occur from the entry of blood into the **kidneys** to the exit of blood from the kidneys. Write the letters A to E in the diagram to show the correct order.

A Wastes such as urea are carried out of the nephron to the bladder, whilst reabsorbed materials leave the kidneys in the renal vein.

Start here

C Useful products are reabsorbed from the nephron and enter the capillaries.

B Small molecules are squeezed into the Bowman's capsule. Large molecules remain in the blood.

D Molecules travel from the Bowman's capsule along the nephron.

E Blood enters the kidney through the renal artery.

Q2 The **blood** entering the kidney contains the following:

ions water proteins sugar urea blood cells

a) List any substances that are usually:

i) filtered out of the blood ....

ii) reabsorbed ....

iii) released in the urine ....

b) Which process is responsible for the **reabsorption** of each substance you have listed above?

....

c) **i)** Name two things from the list above that do **not** enter the Bowman's capsule at all.

....

ii) Explain why these things do not leave the bloodstream.

....

Q3 Three people are tested to see how healthy their kidneys are. Levels of **protein** and **glucose** in their urine are measured. The results are shown in the table.

Which of the three subjects might have kidney damage? Explain how you decided.

....

....

....

Subject	Protein (mg/24 hours)	Glucose (mmol/litre)
1	12	0
2	260	1.0
3	0	0

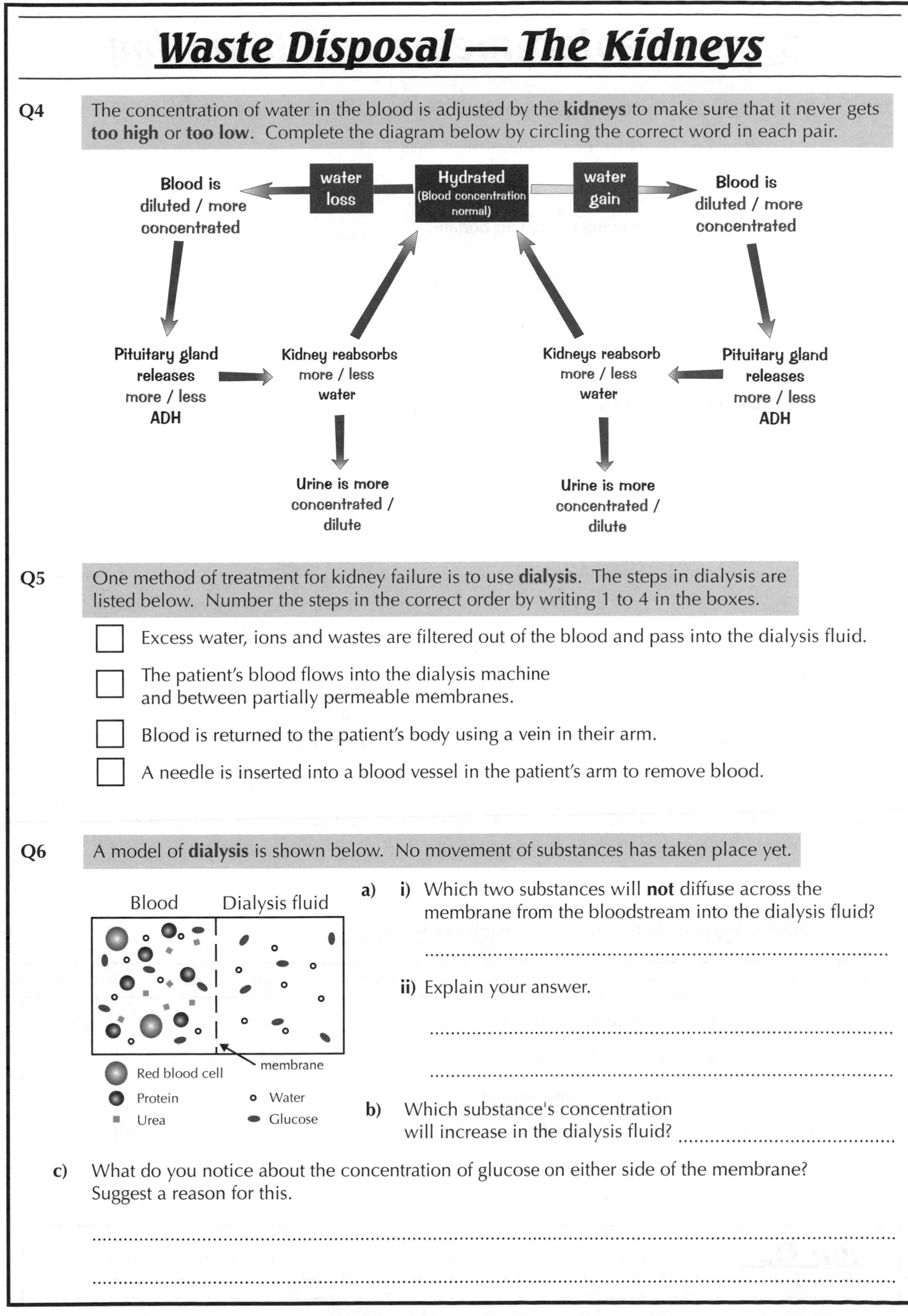

Waste Disposal — The Kidneys

Q4 The concentration of water in the blood is adjusted by the **kidneys** to make sure that it never gets **too high** or **too low**. Complete the diagram below by circling the correct word in each pair.

Q5 One method of treatment for kidney failure is to use **dialysis**. The steps in dialysis are listed below. Number the steps in the correct order by writing 1 to 4 in the boxes.

☐ Excess water, ions and wastes are filtered out of the blood and pass into the dialysis fluid.

☐ The patient's blood flows into the dialysis machine and between partially permeable membranes.

☐ Blood is returned to the patient's body using a vein in their arm.

☐ A needle is inserted into a blood vessel in the patient's arm to remove blood.

Q6 A model of **dialysis** is shown below. No movement of substances has taken place yet.

a) i) Which two substances will **not** diffuse across the membrane from the bloodstream into the dialysis fluid?

....

ii) Explain your answer.

....

....

b) Which substance's concentration will increase in the dialysis fluid? ....

c) What do you notice about the concentration of glucose on either side of the membrane? Suggest a reason for this.

....

....

Organ Replacements and Donation

Q1 Living people can donate organs but to be a **living donor** you must meet certain criteria.

a) List the **three** criteria which living donors must meet.

....

b) Name something that a living donor can donate. ....

Q2 Many people wish to **donate organs** after they die.

a) What can a person do to officially state their wish to donate their organs?

....

b) Who can **object** to the removal of the donor's organs?

....

c) Give **two** possible religious objections to organ donation.

....

....

Q3 Temporary mechanical **replacement organs** can be used to keep people alive.

a) Draw lines to match each mechanical organ with its use.

Heart-lung machine	Filters the blood
Ventilator	Oxygenates and pumps the blood
Kidney dialysis machine	Delivers air to the lungs

b) Give **one** example of when a patient might need a temporary mechanical replacement organ.

....

Q4 Complete the passage by choosing from the words below.

skill	drugs	heart	age	success	rejection	organ	kidney	surgery

Transplants are risky because they involve major .... and .... rates depend on many factors. E.g. the type of .... transplanted is important — a .... is riskier to transplant than a .... The .... of the patient and the .... of the surgeon are also important. Even if all goes well there can be problems caused by .... or by taking immunosuppressive ....

Top Tips: This is quite a challenging topic and there are lots of ethical and religious issues involved. It's important to learn about these because they could easily come up in the exam.

Bones and Cartilage

Q1 Vertebrates have an **internal skeleton** whereas insects have an external one.

a) Give **three** functions of a skeleton.

....

b) Give **two** advantages of an internal skeleton compared to an external one.

....

Q2 The diagram below shows a **long bone**.

a) Name the parts **A**, **B** and **C** on the diagram.

A .... B .... C ....

b) What is the function of part **A**?

....

c) What is made in part **C**?

....

A
B
C

Q3 Complete the passage about **bones** and **cartilage** using the words below.

Some words may be used more than once.

infected cartilage grow calcium phosphorus X-ray bone repair ossification

Bones and cartilage are made of living cells — this allows them to .... and .... themselves, but it also means that they can become .... In the womb, bones start off as .... They turn to bone when .... and .... are deposited as you grow. This process is called .... In people who are still growing, there is a lot of .... present. You can look at how much is present using an .... produces an image, but .... doesn't.

Q4 Elderly people often suffer from **osteoporosis**, which makes their bones more prone to **fractures**.

a) Why are their bones more likely to break?

....

b) Explain why you should try to avoid moving anyone who might have a fracture.

....

....

Joints and Muscles

Q1 Complete the passage below to explain how different **joints** allow different ranges of movement.

socket	hinge	hip	knee	ball	shoulder	elbow	rotate	one

.... joints, such as the .... and ...., allow movement in .... direction only. .... and .... joints, such as the .... and ...., allow movement in many directions — they can also ....

Q2 A damaged hip can be replaced with an **artificial joint** to help walking and to reduce pain. However, there are problems with artificial joints. Give two **disadvantages** of artificial joints.

1. ....

2. ....

Q3 Draw lines to match each part of a **synovial joint** to its function.

Cartilage	Hold the bones together
Synovial membrane	Acts as a shock absorber
Ligaments	Lubricates the joint
Synovial fluid	Produces synovial fluid

Q4 The diagram below shows the arrangement of the **muscles** and **bones** in the arm.

a) Name the parts **A**, **B**, **C**, **D** and **E** on the diagram.

A .... B ....

C .... D ....

E ....

A
B
C
D
E

b) Name the part of the arm that acts as a **pivot**. ....

c) Give the letter of the muscle that is **contracted** in the diagram. ....

d) Give the letter of the muscle that is **relaxed** in the diagram. ....

Top Tips: There are loads of different types of joints but you only need to know about two — hinge, and ball and socket... so get them learnt. Don't forget that muscles usually work in pairs.

Bacteria

Q1 The diagram shows a typical **bacterium**.

a) Name parts A and B on the diagram.

A ....

B ....

b) Give **two** ways in which this cell is different from an animal cell.

1. ....

2. ....

c) Name **one** feature of a typical plant cell that is not seen in bacterial cells.

....

Q2 Draw lines to match up each **part** of a bacterium to its correct **description**.

Part	Description
DNA	helps to stop the cell from bursting
Flagellum	genetic material found in the cytoplasm
Cell wall	helps the cell to move

Q3 Answer the following questions about **bacterial reproduction**.

a) Bacteria reproduce to give clones. What is a clone?

....

b) Explain what happens when bacteria undergo binary fission.

....

Q4 Explain how storing food in a **fridge** can help to stop it **going off** so quickly.

....

....

....

Harmful Microorganisms

Q1 Microbes must **enter the body** before they can cause a disease. Give **four** ways that microorganisms can do this and in each case give an **example** of one **disease** caused by a microbe that uses that route.

When their usual route is cancelled, they have to use a replacement bus service.

1. ....
2. ....
3. ....
4. ....

Q2 Explain what happens to **cause the disease symptoms** once the microbe has entered the body.

....

....

....

Q3 Developing countries often have a **higher incidence of infectious disease** than developed countries do. Explain fully why you think this is.

....

....

....

....

Q4 Disease spreads rapidly after **natural disasters**.

Explain how **damage** to each of the following can lead to ill health:

a) Transport systems ....

....

b) Sewage systems ....

....

c) Electrical supplies ....

....

Microorganisms and Food

Q1 Complete the passage about **yoghurt making** by filling in the gaps using the words below.

cooled ferment flavours clot pasteurised lactic acid bacteria incubated

To make yoghurt, milk is .... to kill off any unwanted microorganisms, then .... Next, a starter culture of .... is added and the mixture is ....
The bacteria .... the lactose sugar into ....
This causes the milk to .... and form yoghurt. .... such as fruit are then sometimes added.

Q2 Number these steps in the manufacture of **soy sauce** to give the correct order.

.... fermentation by *Aspergillus*

.... filtering

.... soy beans and roasted wheat are mixed

.... pasteurisation

.... fermentation by yeast

.... fermentation by *Lactobacillus*

Q3 Some people take **prebiotics** to promote the growth of **'good' bacteria** in the gut.

a) What are prebiotics?

....

b) Why are humans and 'bad' bacteria unable to digest prebiotics?

....

c) Give two natural sources of prebiotics.

1. .... 2. ....

Q4 **Microorganisms** are used in the manufacture of various **additives** and **supplements**. Draw lines to match each of the microorganisms below with the substance it is involved in making.

a) ***Acetobacter*** — **monosodium glutamate**

b) ***Aspergillus niger*** — **low-calorie sweetener**

c) ***Corynebacterium glutamicum*** — **vitamin C supplement**

d) ***Saccharomyces cerevisiae*** — **citric acid**

Top Tips: Remember some bacteria are 'bad' and can cause disease but there are also 'good' bacteria. Everyone has 'good' bacteria in their guts — they're really important for digestion.

Microorganisms and Food

Q5 Scientists did an experiment into the effectiveness of **stanol esters** in lowering people's **blood cholesterol**. They asked two groups of 100 people each to use a special spread instead of butter. Group A's spread was based on vegetable oil. Group B's spread was exactly the same, except that it contained large amounts of stanol esters. The cholesterol levels of each group were measured before the start of the experiment, and again after six months. The results are shown in the table.

	Group A / units	Group B / units
Mean blood cholesterol at start	6.3	6.4
Mean blood cholesterol after 6 mths	6.1	5.5

a) Explain the purpose of Group A.

....

b) Why did the scientists use 100 people in each group?

....

c) Why is it necessary to measure the blood cholesterol before the experiment as well as at the end?

....

d) Explain why it is important that people with high blood cholesterol take steps to lower it.

....

e) Explain how bacteria are involved in making spreads such as that used by group B.

....

....

Q6 Some microorganisms are used as additives in food. Additives are added to foods for different reasons. Draw lines to match each of these **food additives** to its **function**.

a)	**Chymosin**	**sweetener**
b)	**Fructose**	**flavouring**
c)	**MSG**	**preservative**
d)	**Vitamin C**	**clotting agent**

Q7 From the list of **food products** in the box, choose the product or products:

carrageenan	**citric acid**	**fructose**
soy sauce	**stanols**	**yoghurt**

a) used as a preservative. ....

b) that depend on fungi for production. ....

c) that depend on bacteria for production. ....

d) that do not require microorganisms for production. ....

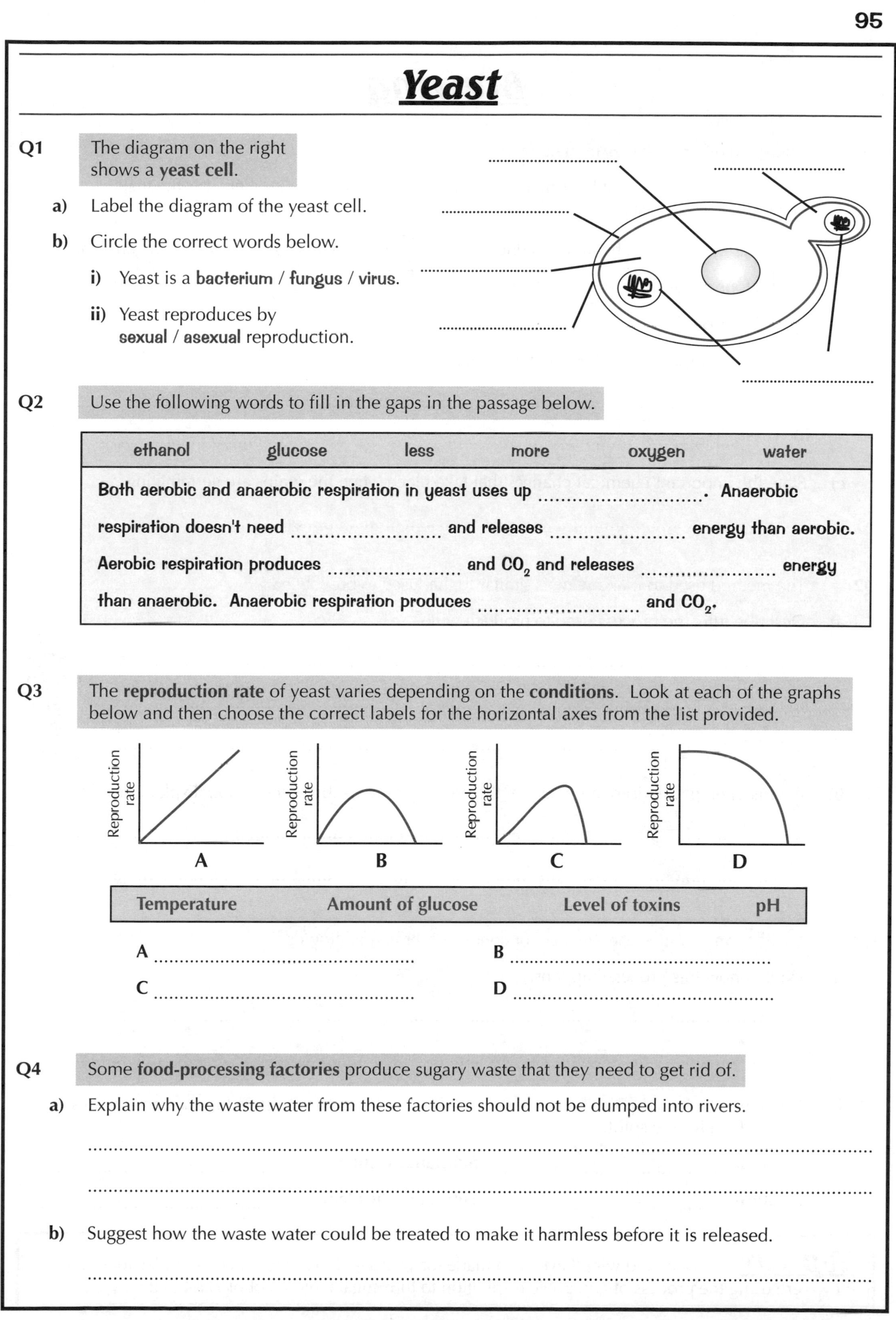

Yeast

Q1 The diagram on the right shows a **yeast cell**.

a) Label the diagram of the yeast cell.

b) Circle the correct words below.

i) Yeast is a **bacterium / fungus / virus**.

ii) Yeast reproduces by **sexual / asexual** reproduction.

Q2 Use the following words to fill in the gaps in the passage below.

ethanol	glucose	less	more	oxygen	water

Both aerobic and anaerobic respiration in yeast uses up .... Anaerobic respiration doesn't need .... and releases .... energy than aerobic. Aerobic respiration produces .... and CO_2 and releases .... energy than anaerobic. Anaerobic respiration produces .... and CO_2.

Q3 The **reproduction rate** of yeast varies depending on the **conditions**. Look at each of the graphs below and then choose the correct labels for the horizontal axes from the list provided.

Temperature	Amount of glucose	Level of toxins	pH

A .... B ....

C .... D ....

Q4 Some **food-processing factories** produce sugary waste that they need to get rid of.

a) Explain why the waste water from these factories should not be dumped into rivers.

....

....

b) Suggest how the waste water could be treated to make it harmless before it is released.

....

Brewing

Q1 There are different stages in making **beer**.

a) Below are some stages of beer making. Number them from 1 to 5, where 1 is the first stage and 5 is the last.

☐ Barley grains are heated and dried.
☐ Barley grains are soaked and germinated.
☐ Hops are added.
☐ Malted grains are mashed with water.
☐ Yeast is added.

b) Explain the purpose of adding the following:

i) hops ....

ii) yeast ....

c) State the important chemical changes that take place when the grains are germinating.

....

Q2 The method used to make **wine** is similar to that used to brew beer.

a) Describe fully the process used to produce wine.

....

....

....

b) Beer is often pasteurised at the end of the brewing process, but wine is not. Explain this difference.

....

....

Q3 **Distillation** can increase the alcohol content of fermented liquids.

a) Explain how this process happens.

....

....

b) Name two types of 'spirit' that are made in this way. In each case, state the substance that is distilled to give the spirit.

1. **Spirit:** .... **Substance distilled:** ....

2. **Spirit:** .... **Substance distilled:** ....

Top Tip: Beer and wine have been made for yonks — long before people understood that they were using the process of anaerobic respiration to turn sugars into alcohol. Not that they cared.

Fuels from Microorganisms

Q1 Use the words below to fill in the gaps and complete the passage about **biogas**.

batch	generator	fermented	heating	turbine	waste

Biogas can be made in a container called a ...., either by continuous production or by .... production. It is made from plant and animal ...., which is .... by microorganisms. Biogas could be used for ...., or even to turn a .... for making electricity.

Q2 Below are some stages in the making of **gasohol**.

- ☐ The product is mixed with petrol.
- ☐ Carbohydrase is added.
- ☐ Starch is extracted from a plant.
- ☐ The mixture is distilled.
- ☐ Yeast is added.

a) Number these stages 1 to 5, where 1 is the first stage and 5 is the last.

b) Explain the purpose of each of these stages:

i) adding the carbohydrase ....

ii) distilling the mixture ....

iii) adding the yeast ....

Q3 Below are some fairly straightforward questions about **biogas**. Great.

a) Name the main components of biogas.

....

b) Name **two** materials that might be used as food for the microorganisms used in producing biogas.

....

c) Suggest an environmental advantage of using biogas (instead of natural gas) to heat a home.

....

Fuels from Microorganisms

Q4 Biogas may be produced in a **batch** or **continuous** generator from waste materials, e.g. animal waste. Circle the correct words below to describe how a batch and continuous generator differ.

a) In a **batch / continuous** generator, waste is usually loaded manually.

b) A **batch / continuous** generator is the best choice for large-scale production.

c) In a continuous generator, waste is added **at intervals / all the time**.

d) In a **batch / continuous** generator, biogas is produced at a steady rate.

Q5 In a village in South America, a **biogas generator** was built.

a) Suggest reasons for the following features of the design:

i) The generator was built some distance away from houses in the village.

....

ii) The generator was built close to fields where animals were grazing.

....

iii) The generator was covered with insulating material.

....

b) Describe **two** possible advantages for the villagers in having a biogas generator like this.

....

....

Q6 The diagram shows a **biogas generator system**.

a) The energy in biogas originally came from the **Sun**. Explain how.

....

....

b) How can biogas power **electrical appliances**?

....

c) Biogas is sometimes described as being '**carbon neutral**'.

i) Explain why biogas is carbon neutral.

....

....

ii) Name **one** other type of fuel which is **not** carbon neutral. ....

Enzymes in Action

Q1 **Enzymes** are used for a variety of purposes.

a) What is an enzyme?

....

....

b) Give three uses of enzymes in the home or in industry.

1. ....

2. ....

3. ....

Q2 **Bacteria** produce enzymes to help break down their food. One example is the **protease enzyme**, which breaks down **protein** into amino acids. Some bacteria live at **high temperatures** and so their enzymes work best at high temperatures too.

a) Some **washing powders** have the protease added to them. Explain why.

....

....

b) Suggest why the protease in some newer powders is extracted from bacteria that live at high temperatures, rather than from bacteria that live at room temperature.

....

....

Bac-Wash

Q3 Explain the role of each of the following enzymes in the **food industry**:

a) Rennet ....

....

b) Invertase ....

....

c) Pectinase ....

....

Top Tip: When you're talking about these reactions, remember that the enzymes only catalyse the reactions. Don't make it sound like they actually take part in them, or you'll look a fool. Like a fooool!

Enzymes in Action

Q4 There are several ways to test for the presence of **sugar** in a solution. Two of these methods are described below.

The Benedict's test
This involves heating the solution with blue Benedict's reagent. There is a gradual change from blue to orange if sugar is present.
The orange substance formed is solid and eventually settles out to form a layer. Lots of different sugars will produce this reaction.

Reagent strips
The tip of the strip has a small pad with enzymes in it. The enzymes catalyse a reaction involving glucose and this makes the pad change colour. Only glucose results in this change. Different colours are produced for different concentrations.

a) Suggest three reasons why the reagent strip is a better method than the Benedict's test for testing the concentration of glucose in a solution.

1. ....

2. ....

3. ....

b) Give one way in which these reagent strips are being used in medicine.

....

Q5 Enzymes can be **immobilised** using various methods.

a) Give one way that enzymes can be immobilised.

....

b) Give three advantages of using immobilised enzymes.

1. ....

2. ....

3. ....

Q6 Some kinds of **dairy product** are treated to remove the **lactose** (milk sugar).

a) Explain why this is done.

....

....

b) Explain how immobilised enzymes are used to produce lactose-free milk.

....

....

Genetically Modifying Plants

Q1 Some stages in the production of a **herbicide-resistant maize plant** are listed below. Put the stages into the correct order.

A The herbicide-resistance gene is inserted into *Agrobacterium tumefaciens*.

B Infected cells from maize are grown in a medium containing herbicide.

C The gene that makes a wild corn plant resistant to herbicide is identified.

D *Agrobacterium tumefaciens* is allowed to infect a maize plant.

E The herbicide-resistance gene is cut out from a wild corn plant.

Order: ....

Q2 A crop plant had been genetically modified to make it **resistant to herbicides**. Some people were **concerned** that, as a result, wild grasses growing nearby might also become resistant to herbicides. Scientists decided to check whether this had happened.

The scientists sprayed herbicide onto 100 plants in an area next to the GM crop, and onto 100 plants in a second area far away from the GM crop. The results are shown in the table.

Number of grass plants dying after spraying	
In area next to GM crop	In area far away from GM crop
83	85

a) Explain the reason for testing a group of plants that had not been growing near the GM crop.

....

b) How could the scientists have made the results of this experiment **more reliable**?

....

c) The scientists decided that there was no significant difference between the two groups of plants. Explain whether you agree or disagree with this conclusion.

....

....

d) If the scientists are right in their conclusion, does this prove that the concerns about genes for resistance spreading are unfounded? Explain your answer.

....

....

....

e) If wild grasses become resistant to herbicides, what problems might this cause?

....

....

Developing New Treatments

Q1 Use the words provided below to fill in the blanks in the passage.

design **genomics** **medicine** **predispose** **prevent**

The study of all the genes in an organism is called ...., and it could be useful in ...., to develop treatments. Some people have genes which .... them to certain diseases. Identifying these genes in a person may help to .... the diseases ever developing. One day, it may also help doctors to .... specific drugs which suit an individual patient's particular genetic make-up.

Q2 The plant ***Artemisia annua*** has been used in traditional Chinese **medicine** for a long time.

a) A new drug was made by extracting the chemical which gave the medicinal effect. What name is given to such a chemical? ....

b) The new drug is called artemisinin. What is it used to treat? ....

Q3 The graph shows the amount of **money** spent on buying **drugs** by the health services of two countries with similar population sizes.

a) How has spending by country A changed between 1990 and 2006?

....

....

b) Give **two** possible reasons for this trend.

....

c) Suggest why the spending by country **B** does not show a similar trend.

....

d) Explain how abandoning the patent system for new drugs might help country B.

....

....

e) Explain why abandoning the patent system might **not** be a good idea for people in either country.

....

....

Instinctive and Learned Behaviour

Q1 Complete the passage by choosing from the words below.

genes moisture environment learned light

Most behaviours seen in animals are due to both inherited and .... factors.
Inherited aspects of behaviour depend on the animal's ....
An example of inherited behaviour is the negative phototaxis of earthworms, where they move away from ....

Q2 Match up the aspects of **human behaviour** to show whether they are **instinctive** or **learned**.

Playing football | instinctive | Language

Salivating | learned | Sneezing

Q3 A student was studying the behaviour of **birds** on a **bird table**. Each day the student provided small pieces of cheese, some nuts hanging from the table on lengths of string, and some corn.

Below are some of the observations made by the student:

1. **Robins took the cheese, but ignored the nuts and corn.**
2. **Pigeons took the corn, but ignored the cheese. They expressed interest in the hanging nuts, but weren't able to get at them.**
3. **Great tits took some of the cheese, and managed to hang from the strings to take the nuts.**

Occasionally a single crow visited the bird table. It initially took the cheese and corn, and watched the great tits. After the third week the student recorded that the crow also began to take the nuts. It was able to get at them by hanging from the string.

a) Give one example of a behaviour mentioned that seems to be **instinctive**, and explain your answer.

....

....

b) Give one example of a behaviour mentioned that is **learned**, giving a reason for your answer.

....

....

c) The student wanted to study how experiences in early life effect bird development. They kept a bird in isolation from a young age. What effect would you expect this to have on its song?

....

Top Tips: Animals are born with all the nerve pathways they need for instinctive behaviours already connected. The nerve pathways needed for learned behaviours develop with experience.

Instinctive and Learned Behaviour

Q4 **Habituation** is an important part of the learning process in young animals.

a) Explain the term **habituation**.

....

....

b) Explain why habituation is beneficial to animals.

....

Q5 Explain what a '**Skinner box**' is and describe how it can be used to study animal behaviour.

....

....

....

Q6 **Conditioning** is a type of **learned behaviour**.

a) Explain the difference between '**classical** conditioning' and '**operant** conditioning'.

....

....

....

....

b) Describe an example of:

i) classical conditioning ....

ii) operant conditioning ....

Q7 **Guide dogs** for the blind undergo a period of intensive **training**. One part of this training involves teaching the dogs to stop at roadsides and wait for commands.

a) Suggest one form of operant conditioning that could be used to ensure that the dog learns to stop at the roadside and wait for a command.

....

b) Explain why operant conditioning involving rewards is preferable to operant conditioning involving punishments during animal training.

....

Social Behaviour and Communication

Q1 List three reasons why animals **communicate** with one another.

1. ....

2. ....

3. ....

Q2 Below is a list of different **types** of communication used by different kinds of animals. In each case, suggest a **reason** for the communication.

a) A female moth releases a pheromone into the air.

....

b) A honey bee does a 'waggle dance' in the hive.

....

c) A dog rolls onto its back.

....

Q3 **Peafowl** are large birds related to pheasants. Male peafowl, called **peacocks**, have long coloured feathers that project beyond the tail. Peafowl live naturally in India, where they are sometimes preyed on by tigers.

a) Suggest a possible **advantage** of the long feathers to the peacock.

....

b) Suggest a possible **disadvantage** of having these long feathers.

....

c) Female peafowl, called peahens, are dull-coloured in comparison. Suggest why.

....

....

Q4 **Language** is the most obvious form of **human communication**, but there are others.

a) Give three methods of **non-verbal** communication between humans.

....

b) What is a possible advantage of verbal communication over non-verbal communication?

....

Social Behaviour and Communication

Q5 The **chiffchaff** and the **willow warbler** are two related species of woodland birds. They are both green-brown in colour and spend much of their time among the foliage of trees.

a) Suggest why these birds attract mates using song rather than visual signals.

....

b) The song of the chiffchaff sounds very different from the song of a willow warbler. Explain why this is necessary.

....

....

Q6 **Communication** can happen in many different ways.

a) Humans communicate using **speech** and birds communicate using **song**. In what ways are these two forms of communication different?

....

....

b) Humans and many other mammals also use **facial expressions** to communicate. Would you expect a panda bear to understand a human frown? Explain your answer.

....

Q7 When confronted with a **mirror**, a **dog** may look behind the mirror in an attempt to find the 'other animal' presented to it. Suggest how this kind of reaction makes the behaviour of the dog fundamentally **different** from the behaviour of a typical human being.

....

....

Q8 It's thought that humans are more **self-aware** than other animals.

a) Explain what is meant by the term **self-awareness**.

....

....

....

b) Explain why it is difficult to know whether other animals have 'self-awareness'.

....

Feeding Behaviours

Q1 Choose words from the list below to complete the following passage.

carnivores	vitamin A	herds	predators	amino acids	prey	herbivores	packs

Sheep, cows, horses and rabbits are ...., which means that they feed on plant material. Many feed in groups called .... This makes it more likely that at least a few individuals will be able to spot ....
A problem with a herbivorous diet is that it can be low in certain kinds of nutrients, such as ...., so herbivores have to spend a lot of time feeding.

Q2 Tick the boxes to show which of the following statements are most likely to apply to **herbivores**, and which to **carnivores**.

		herbivores	carnivores
a)	They have strong horns for defence.	☐	☐
b)	They eat food that is high in protein.	☐	☐
c)	They can go for several days without feeding.	☐	☐
d)	They form groups with others of the same species for safety.	☐	☐
e)	They form groups with others of the same species to get food.	☐	☐

Q3 **Wolves** work in packs when hunting large animals such as **reindeer**.

a) Suggest why it is important for wolves to cooperate in this way when they hunt reindeer.

....

b) Reindeer are usually found in **herds**. Give two reasons why this reduces the chances of any one particular reindeer being caught by a wolf pack.

....

....

....

c) Explain why large herds of herbivores need to move around frequently.

....

d) Wolves usually hunt **individually** for smaller prey such as rabbits. Suggest why.

....

Feeding Behaviours

Q4 Experiments were carried out to investigate the '**begging response**' in young **herring gulls**. The young peck at the bill of the parent to stimulate it to regurgitate fish, which the young then swallow. This behaviour occurs soon after the young hatch. Scientists presented young herring gulls with a series of **cardboard models** of a parent gull's head. The results of the study are shown below. Real adult herring gulls have a **white head**, with a **yellow bill** and a **red spot** near the tip.

Model	White head, grey bill, no spot	White head, grey bill, red spot	White head, yellow bill, red spot	Pointed red stick with three white bands
No. of pecks by young	5	39	42	50

a) Describe what these experiments demonstrate about what stimulates the begging response in young herring gulls. Explain your answer.

....

....

....

b) Is the begging response instinctive or learned? Give a reason for your answer.

....

....

c) It's thought that parent birds are also stimulated to regurgitate food by the wide open, brightly coloured mouths of chicks begging for food. Describe how you could test whether it's the **colour** or the **size** of a chick's open mouth (or both) which stimulates the parent to regurgitate food.

....

....

Q5 Some animals use **tools** to get food.

a) Give two examples of animals using tools.

1. ....

2. ....

b) Some species of vulture break open bones to reach the nutritious marrow inside by dropping the bones onto rocks. Would you class the rocks as tools in this case? Explain your answer.

....

....

Reproductive Behaviours

Q1 Draw lines to match up each animal below with the most likely way in which a **male** of that species would **attract a mate**.

mandrill monkey
red deer
frog
moth

display aggression to other males
mating call
pheromone
display brightly coloured parts of body

Q2 Explain what is meant by the following terms:

a) monogamy

....

b) harem

....

c) courtship

....

Q3 In many species of **bird**, **both** parents play a role in incubating eggs and feeding the young once the eggs have hatched.

a) State an advantage of this shared responsibility for:

i) the young. ....

....

ii) the parents. ....

....

b) **Birds of paradise** differ in that the females have sole responsibility for looking after the young. These birds live on the island of New Guinea, where there are few predators.

Suggest a possible link between the reproductive behaviour of birds of paradise and the fact that there are few predators in their habitat.

....

....

....

Reproductive Behaviours

Q4 In most species, **males compete** to win the right to mate with females. Methods used vary from bringing the female gifts of food, to fighting off the other males. However, it is the **opposite** way around in **seahorses** — females compete for the attention of males. Seahorses are also unusual in that the female lays her eggs in the male's pouch and **he** is then '**pregnant**' with the young and eventually gives birth to them.

a) Explain fully why males usually compete for females, and why this is not the case in seahorses.

....

....

....

....

....

b) Why is it important for most animals that females don't mate with a male of a closely related species by mistake?

....

....

Q5 Some animals **care for their young** for long periods, while others provide **no parental care** at all.

a) Name three animals that care for their young, and three that do not.

Care: 1. ....

2. ....

3. ....

Don't care: 1. ....

2. ....

3. ....

b) Give three ways in which animals may care for their young.

1. ....

2. ....

3. ....

Top Tips: The male bowerbird impresses females by constructing an elaborate mound of earth decorated lavishly with shells, leaves, feathers and flowers, which he spends hours carefully arranging.

Living in Soil

Q1 An area of garden soil was **waterlogged** after a nearby water pipe burst. **Earthworms** were then observed coming to the surface of the waterlogged area more often. The soil stayed waterlogged for a week, during which time the **plants** growing in the soil withered and died.

a) Suggest an explanation for the behaviour of the earthworms.

....

....

b) Suggest how the condition of the soil might be affected over time if the earthworm population decreased.

....

....

Q2 **Bacteria** have an important role in **recycling** the nutrients from dead material.

a) Describe the roles of different types of bacteria in converting the nitrogen in dead material into a form that can be taken up by plants.

....

....

....

....

b) Suggest one way that a farmer could increase the levels of nitrate in his soil.

....

Q3 The trees in **tropical rainforests** have quite **shallow roots** compared with the trees found in Britain. The roots of many tropical trees are infected with certain kinds of **moulds** that have, in turn, many microscopic connections with **dead leaves** on the forest floor. As the moulds break down the dead leaves, many of the nutrients pass **directly** into the roots without being released into the **soil**.

Using the information above:

a) Suggest why the levels of nitrates are lower in rainforest soils compared with British forest soils.

....

b) Suggest why the roots of many tropical trees are shallow.

....

....

Living in Water

Q1 Life in water and life on land are very different.
Give two **advantages** and two **disadvantages** of living in water.

....

....

....

....

Q2 Most species of crab, such as **spider crabs**, live in **water**. Other species, such as **robber crabs**, live on **land** and will actually drown if dropped into water.

Which species of crab — the spider crab or the robber crab — is likely to be more tolerant of changes in temperature? Give a reason for your answer.

....

....

Q3 Many **plants** that grow on land have **tough woody parts**.
Suggest why few plants found in **water** have woody parts.

Woody parts provide the plant with support.

....

....

Q4 **Amoebas** are single-celled organisms that are found in **freshwater**.

a) Explain why amoebas tend to absorb too much water.

....

....

b) Explain the role of the **contractile vacuole** in preventing amoeba cells bursting.

....

....

c) Related species of single-celled organism living in the sea don't have a vacuole. Suggest why not.

....

....

Living in Water

Q5 There are **two** types of plankton — **phytoplankton** and **zooplankton**.

a) Describe what phytoplankton and zooplankton are.

....

....

b) Explain the importance of phytoplankton in aquatic food webs.

....

Q6 **Algal blooms** can arise in ponds, making the water go green and murky.

a) Explain why algal blooms usually occur in the summer and not the winter.

....

....

b) What effect will an algal bloom have on the zooplankton in the pond? Explain your answer.

....

Q7 **Amphibians** and **fish** have different methods of gaseous exchange from mammals.

a) i) How do adult **amphibians** exchange oxygen and carbon dioxide?

....

ii) How and why does this restrict where they can live?

....

....

b) i) Name the structure that fish use for gaseous exchange, and give one way in which this structure is adapted for its function.

....

....

ii) Why are fish unable to breathe out of water?

....

....

Top Tips: Plankton may seem tiny and insignificant, but they support entire aquatic food webs. Some massive animals, like a lot of whales, feed on them almost exclusively. They're very important.

Human Evolution and Development

Q1 Suggest how each of the following has contributed to the **success** of the **human species**:

a) Tool use ....

....

b) Living in large, complex societies ....

....

Q2 Below are some of the key events in the **evolution** of **human behaviour**.

2.5 million years ago — earliest use of stone tools.

50 000 years ago — hunter-gatherer society.

10 000 years ago — farming begins in some parts of the world.

a) Explain what is meant by 'hunter-gatherer society'.

....

....

b) Explain how the use of tools would have been beneficial in such a society.

....

....

c) Suggest one advantage of a farming system over a hunter-gatherer system.

....

Q3 Humans have **domesticated** a range of animal species.

a) Dogs were one of the first species to become domesticated — about 14 000 years ago. Explain the advantage to the humans living at this time of domesticating dogs.

....

....

b) Choose reasons from the list below for domesticating the following animals.

A For hide **B** Farming the land **C** For food **D** For travelling **E** Carrying heavy loads

Horse .... **Cattle** ....

c) Give one example of how selective breeding has been used to modify the characteristics of a domesticated animal.

....

Human Behaviour Towards Animals

Q1 Name **one animal** that is used for each purpose given below.

a) Providing wool ....

b) Hunted for sport ....

c) Racing ....

Q2 Many different species of animal are kept in **zoos**. Some people think that this is **necessary**, but others think it is **cruel**. Outline the reasons for and against keeping animals in zoos.

....

....

....

....

Q3 Animals are often used to test **drugs** before they are released for human use.

a) Explain why many people think this is necessary.

....

b) Suggest one reason why some people feel it is cruel to use animals in this way.

....

c) Give one other way in which animals could be of use in medicine.

....

Q4 Some people feel that animals should have the **same rights** as humans.

a) Animals are used for entertainment in some circuses.

i) Give one reason why some people might be against this.

....

ii) Explain how a circus owner might justify this use of the animals.

....

b) Animals are also bred for food. They are often intensively farmed. Explain:

i) why some people think that this is wrong.

....

ii) how this use of animals could be justified.

....

Answers

Section One — Nerves and Hormones

Section One — Nerves and Hormones

Page 1 — The Nervous System

Q1 neurones, electrical, effectors, glands, sensory/ motor, motor/sensory

Q2 a) Chemical receptor. **Tongue** underlined.
b) Chemical receptor. **Nose** underlined.
c) Light receptor. **Eyes** underlined.
d) Sound receptor. **Ears** underlined.

Q3 a) central nervous system
b) brain and spinal cord
c) i) sensory neurone
ii) motor neurone

Q4 The information from the receptors in the toe can't complete its normal path through the spinal cord to the brain.

Page 2 — Reflexes

Q1 The response happens without you having to take time thinking about it.

Q2 The set of nerve cells that carry information from receptor to effector, via the CNS, in an automatic reaction.

Q3 a) i) sensory neurone
ii) relay neurone
iii) motor neurone
b) i) As an electrical signal.
ii) As a chemical signal.
c) effector
d) i) synapses
ii) The signal is transferred across the gap by chemicals, which are released when the impulse arrives at one side of the gap. The chemicals diffuse across it and trigger a new impulse in the neurone on the other side.

Page 3 — The Eye

Q1

iris
cornea
lens
retina
optic nerve

Q2 a) Eye A. The pupil is smaller in this diagram to stop too much light entering the eye and damaging it.
b) Reflex responses happen very quickly, so the eye can respond to changes in light intensity as soon as possible. This helps it adjust quickly to dimmer light, and stops it being damaged by sudden bright lights.
c) binocular vision

Q3 relax, tighten, thin, near, flexibility

Page 4 — Hormones

Q1 chemical, glands, blood, target

Q2 Adrenaline. It is called the 'fight or flight' hormone because it prepares the body for action (usually either defending yourself or running away) when you get a shock.

Q3 a) pancreas
b) FSH
c) sugar
d) blood
e) testosterone

Q4 Responses that are due to hormones generally happen more slowly and last longer than those that are due to the nervous system. Nerves act on a very precise area, while hormones travel all over the body and can affect more than one area at once.

Q5

HORMONE	SITE OF PRODUCTION	ACTION
ADH	**Pituitary gland**	**Controls water content**
Oestrogen	**Ovaries**	Helps control the menstrual cycle
Testosterone	**Testes**	**Promotes all male secondary sexual characteristics at puberty.**

Page 5 — Puberty and the Menstrual Cycle

Q1 a) Any two of, e.g. facial hair / increased body hair / deepening of voice / enlargement of the penis and testicles / sperm production / development of muscles.
b) Any two of, e.g. hips widen / breasts develop / periods start / pubic hair grows.

Q2 a) A — LH/Luteinising hormone
B — FSH/Follicle stimulating hormone
C — Oestrogen
D — Progesterone

b) and c)

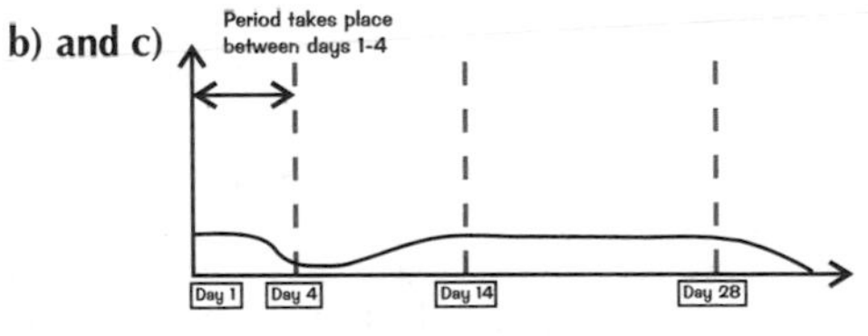

Q3 a) FSH — pituitary gland
oestrogen — ovaries
LH — pituitary gland
progesterone — ovaries
b) Causes pituitary gland to produce LH. Inhibits further production of FSH. Causes the lining of the uterus to thicken and grow.

Q4 a) So that if an egg is fertilised the resulting embryo can attach to the uterus wall to obtain nourishment.
b) The lining starts to break down and is shed at the beginning of the next cycle.

Section Two — Diet and Health

Page 6 — Controlling Fertility

Q1 a) FSH
b) FSH, oestrogen, pituitary gland, LH, ovary, egg
Q2 Reducing fertility — any two of, e.g. not 100% effective / can cause side effects like nausea/ headaches/irregular periods / doesn't protect against STIs.
Increasing fertility — e.g. doesn't always work / can result in multiple births.
Q3 Oestrogen is taken every day to give high levels in the blood. This inhibits the release of FSH by the pituitary gland. Eventually egg development in the ovaries stops and no more are released.
Q4 a) Hormones (usually FSH) are given to the woman to increase the number of eggs that develop. The eggs are collected from her ovaries and fertilised outside the body. They're allowed to develop into embryos, and several are then placed back in her uterus. Oestrogen and progesterone are often given to increase the chance of the embryos implanting.
b) **Advantage** — any one of, e.g. It allows infertile couples to have children. / It allows screening of embryos for genetic defects.
Disadvantage — any one of, e.g. There can be reactions to the hormones, e.g. vomiting, dehydration, abdominal pain. / There may be an increased risk of some types of cancer. / The process can be expensive, there's no guarantee that it will work and it may result in multiple births. / It could open the way to ethically questionable embryo selection practices.

Pages 7-8 — Homeostasis

Q1 The maintenance of a constant internal environment.
Q2 If your blood glucose level is too high, insulin is secreted, which causes glucose to be removed from the blood.
Q3 hot, sweat a lot, less, dark, less, concentrated
Q4 a) More water. The exercise will increase their temperature, so they will have to sweat more to cool down.
b) More water. The exercise will make them breathe faster, so more water will be lost via the lungs.
c) Less water. More will be lost as sweat and in the breath, so to balance this the kidneys will give out less water in the urine.
Q5 a) The enzymes controlling all the reactions in the human body don't work as well if the temperature varies too much (and are denatured/destroyed if they get too hot).
b) the brain
c) You sweat. More blood flows near the skin. Hairs lie flat to allow easier heat loss.
Q6 Ronald's kidneys remove salt from his blood. Ronald gets rid of salt in his urine.
Q7 20th July was hotter so she sweated more. Because more water was lost through her skin, less was lost in her urine to balance this. Her urine had to be more concentrated, so the ion concentration was higher.

Page 9 — Controlling Blood Sugar

Q1 a) From digested food and drink.
b) Liver and pancreas
c) insulin
Q2 a) A condition where the pancreas doesn't produce enough insulin.
b) E.g. using a glucose-monitoring device that checks a drop of their blood.
Q3 a) Any two of, e.g. diet / exercise / by injecting insulin.
b) The insulin causes her liver to store glucose from her blood. Eventually her blood sugar levels get so low that she faints (because there isn't enough glucose to release energy for her brain cells to work properly).
c) i) Without insulin none of the sugar in Paul's blood is stored. As Paul digested the food his blood sugar level rose and rose until he eventually lost consciousness.
ii) Paul would be given insulin to bring his blood sugar level down.

Section Two — Diet and Health

Page 10 — Eating Healthily

Q1

Nutrient	Function
Carbohydrates	Provide energy.
Proteins	Growth and repair of tissues.
Fats	Provide energy and act as an energy store.

Q2 Carbohydrate — Simple sugars
Protein — Amino acids
Fat — Glycerol, Fatty acids
Q3 a) Essential amino acids are amino acids which can't be made by the body.
b) From your food.
Q4 Wendy needs more carbohydrate and protein in her diet because she is more physically active. She needs more protein for muscle development and more carbohydrate for energy.
Q5 The average man is bigger than the average woman and so needs more energy for metabolic reactions and movement. The average man also has more muscle, which needs more energy than fatty tissue.

Page 11 — Diet Problems

Q1 a) $((5 + 6) \div 100) \times 100 = \mathbf{11\%}$
b) Women, because less than half the obese men interviewed knew that they were obese, whereas over half the obese women interviewed were aware of it.
c) Any one of, e.g. heart disease / cancers / diabetes.
Q2 a) Because foods that contain high levels of protein are often too expensive for, or not available to, the poorest people.
b) $0.75 \times 75 = \mathbf{56\ g}$

Section Two — Diet and Health

Q3 a) People with anorexia starve themselves, whereas people with bulimia binge-eat and then make themselves sick.

b) Any three from, e.g. liver failure / kidney failure / heart attacks / muscle wastage / low blood pressure and mineral deficiencies.

Q4 a) BMI = 76 ÷ $(1.7)^2$ = 76 ÷ 2.89 = **26.3**

b) overweight

c) Daniel may have a higher than normal proportion of muscle in his body and muscle weighs more than fat, so he might not be overweight in the usual sense.

Page 12 — Cholesterol and Salt

Q1 a) Something that increases your chance of suffering from heart disease.

b) It increases the risk of high blood pressure/ hypertension.

Q2 a) Not necessarily. There's lots of salt in many processed foods, like breakfast cereals, sauces, soups, crisps, etc.

b) 0.5 × 2.5 = **1.25 g**

Q3 a) Fats attached to proteins.

b) LDL — low density lipoprotein and HDL — high density lipoprotein.

Q4 a) Fatty deposits of cholesterol can block blood vessels. If the flow of blood to the heart muscle is reduced it can be starved of oxygen, causing a heart attack.

b) the liver

Page 13 — Health Claims

Q1 Report B is likely to be more reliable as it was published in a journal, and used a large sample size.

Q2 a) Headlines 2 and 3. They refer to studies based on large samples.

b) E.g. Just because a diet works for one person, it doesn't mean it will work for everyone. / Celebrities often have personal trainers, nutritionists, chefs etc. to help get them in shape, as well as the diet. / They might only be promoting a diet because they're being paid to.

Q3 Some studies have shown a link, but others have not. Results can often be interpreted in different ways, depending on what the researcher wants to show (and who's funding the study).

Q4 a) As a control group.

b) Statins help to lower cholesterol in patients with high levels when combined with lifestyle changes.

Page 14 — Drugs

Q1 a) Drugs are substances which alter the way your body works / alter chemical reactions in your body.

b) Physical addiction means that the body has a physical need for the drug, and will suffer withdrawal symptoms if it's not taken.

c) The body gets used to the drug and higher doses are needed to produce the same effect.

Q2 a) Paul, because he supplied the drugs and this is usually punished more severely than just using the drug is.

b) Janice, because the drug she used (cannabis) is a class C drug, not class B (like the amphetamines Paul and Duncan took).

Q3 depressants, alcohol/solvents, solvents/alcohol, decrease, reactions, judgement, stimulants, nicotine/ecstasy, ecstasy/nicotine, increasing

Q4 Any two of, e.g. unprotected sex / sharing needles / driving under the influence / other accidents / getting into fights.

Page 15 — Drug Testing

Q1
1. Computer models simulate a response to the drug.
2. Drug is tested on human tissue.
3. Drug is tested on live animals.
4. Human volunteers are used to test the drug.

Q2 a) To check whether the drugs have any unknown side effects, and to make sure they work effectively in humans.

b) The whole organism must be used to check the effect of the drug on the whole body.

Q3 a) A placebo is a substance that looks like a drug being tested but contains no drug. It's used so that patients aren't subconsciously influenced to feel better or worse by knowing whether or not they're being treated.

b) A double blind trial is one where neither the scientist doing the test nor the patient knows who is getting the drug and who is getting the placebo.

Q4 a) For use as a sleeping pill.

b) It relieved morning sickness, but it also crossed the placenta and stunted the growth of the foetus's limbs.

c) It is used to treat leprosy, AIDS and certain cancers.

Page 16 — Smoking and Alcohol

Q1 a) 3 + 3 = 6 units.

b) No. 1 unit is at least 20 mg per 100 cm^3. So six units is 120 mg per 100 cm^3. This is more than the legal limit of 80 mg per 100 cm^3.

c) Because your reactions are slower / your coordination is poor / your judgement is impaired.

Q2 a) Any two of, the number of male smokers aged 35-54 has been decreasing since 1950. / The number of female smokers aged 35-54 rose between 1950 and 1970, but then it also began to decrease. / The number of male smokers aged 35-54 has been consistently greater than the number of female smokers aged 35-54.

b) i) The cilia in the lungs and windpipe are damaged by the tar in cigarettes. This means they can't catch as much bacteria before they enter the lungs. The bacteria can cause infections.

ii) Cigarette smoke contains carcinogens.

Q3 a) Because so many more people use them.

Section Two — Diet and Health

b) Any two of, e.g. the NHS spends large amounts each year treating patients with smoking- or drinking-related problems. / The cost to businesses of people missing work due to smoking- or drinking-related problems. / The cost of cleaning up the streets, police time, damage to people and property.

Page 17 — Solvents and Painkillers

Q1 depressants, slowing down, brain damage, irritate, respiratory

Q2 a) She has taken more than the recommended dose. She has taken them after drinking, increasing the chance of an overdose.

b) Mild to moderate pain, fever.

Q3 a) opium, morphine, codeine, heroin

b) Because they are very powerful and highly addictive.

Q4 a) Opiates act on the brain to stop it sensing pain and interfere with the mechanism by which 'pain-sensing' nerve cells transmit messages.

b) Chemicals which cause swelling and sensitise the endings of nerves that register pain.

Page 18 — Causes of Disease

Q1 celled, damaging, toxins, cells, DNA, copies, bursts, damage

Q2 a) A disease-causing microorganism.

b) E.g. bacteria / protozoa / fungi / viruses.

c) They live off the host and provide nothing in return.

Q3 Diabetes — caused by lack of insulin production
Scurvy — caused by a vitamin C deficiency
Anaemia — caused by an iron deficiency
Haemophilia — a genetic disorder

Q4 a) In a benign tumour the cancerous cells do not spread to other sites in the body, but in malignant tumours they do.

b) Any two of, e.g. not smoking / wearing sunscreen (and other safe-sun precautions) / a healthy diet / maintaining a healthy body weight.

Page 19 — The Body's Defence Systems

Q1 a) E.g. the eyes produce lysozyme which kills bacteria. / Stomach acid kills bacteria.

b) E.g. the respiratory system is lined with mucus and cilia which catch bacteria and dust. / The skin provides a physical barrier against microbes.

Q2 a) The cells involved attack all pathogens.

b) phagocytes

Q3 Blood flow to the infected area is increased and plasma leaks into the damaged tissue. This allows white blood cells that can fight the infection to get to the site.

Q4 a) Any one of, antibodies / antitoxins

b) After an infection some white blood cells that can produce antibodies against the pathogen remain in the body. They can reproduce rapidly if they detect the same pathogen again and so fight the infection quickly.

Page 20 — Vaccinations

Q1 a) i) True
ii) True
iii) False
iv) False

b) antibodies, another organism, permanent, temporary

c) Dead and inactive microorganisms are harmless but the body will still produce antibodies to attack them.

Q2 a) Fewer people catch diseases like polio, whooping cough, measles, rubella, mumps and tetanus. These diseases spread more slowly amongst those that do catch them than they did in the past.

b) smallpox

c) Any two of, e.g. Some people do not become immune after vaccination. / Others can experience a bad reaction, e.g. swelling at the injection site. / In rare cases there could even be a serious reaction such as seizures.

Q3 The risk of catching the diseases decreases if more children are immunised. The diseases themselves can sometimes cause more serious complications.

Page 21 — Treating Disease — Past and Future

Q1 Semmelweiss asked all the doctors to wash their hands using antiseptic solution when entering his ward. This killed bacteria on their hands and stopped them from spreading infections to their next patients.

Q2 a) Colds are usually caused by a virus and antibiotics don't kill viruses.

b) Viruses reproduce using your body cells. It's hard to develop drugs which destroy the virus without destroying the cells as well.

Q3 Although most of the bacteria causing Jay's infection would have been killed, a few that were resistant to the antibiotic might survive. These bacteria could reproduce and eventually make a resistant strain.

Q4 a) Natural selection favours bacteria with resistance to antibiotics. Antibiotics can't kill bacteria that are resistant to them, making diseases harder to treat.

b) If a microorganism evolves and changes so it is different from the vaccination, the immune system won't recognise it any more and won't be prepared for an infection.

Section Three — Adaptation and Evolution

Section Three — Adaptation and Evolution

Page 22 — Adaptation

Q1 a) Any two of, fox B is white while fox A is darker in colour. / Fox A has much bigger ears. / Fox B has a more rounded shape than fox A. / Fox B has a bigger surface area than fox A.

b) i) Fox A — desert.

ii) Fox B — Arctic region.

c) To match part a) — any two of,
The different colours of the two foxes help to camouflage them — the white fox B is hard to see against the snow, while fox A blends into the sand. / Large ears give a big surface area for fox A to lose heat from, while small ears help fox B conserve heat. / A rounded shape/smaller surface area means there is a smaller surface over which fox B can lose heat, while fox A has a large body surface for heat loss.

Q2 water, small, concentrated, sweat, nocturnal

Q3 a) In the desert.

b) i) The cactus has spines instead of leaves because the small surface area gives less surface for water to evaporate from. / The spines help to protect the cactus from being eaten by animals.

ii) The cactus has a thick, fleshy stem where it can store water. / The stem can photosynthesise.

iii) The cactus has shallow but very extensive roots, so it can take in as much water as possible when it rains. / The cactus has very long roots that can reach water far below the surface.

Page 23 — Classification

Q1 a) i) kingdom

ii) genus

iii) species

b) A group of closely-related organisms that can breed to produce fertile offspring.

Q2

	Plant	Animal
Travels to new places		✓
Hunts for food		✓
Fixed to the ground	✓	
Compact body		✓

Q3 a) Vertebrates have an internal skeleton with a backbone and invertebrates don't (although they may have an exoskeleton).

b) Frog — Amphibian — Moist, permeable skin, lay eggs in water
Horse — Mammal — Produce live young, produce milk
Snake — Reptile — Dry, scaly skin, lay leathery eggs on land
Sparrow — Bird — Feathers, most fly, mouth adapted into beak
Herring — Fish — Scales and fins, live and lay eggs in water

Q4 a) It shows features of more than one vertebrate class. / It shows features of both birds and reptiles.

b) Long bony tail, claws and sharp teeth.

c) Laid eggs. Archaeopteryx has features of a reptile and a bird — both of these classes lay eggs.

Page 24 — Ecosystems and Species

Q1 a) In natural ecosystems humans don't try to control the processes, but in artificial ecosystems humans promote some organisms and eliminate others.

b) E.g. farms, fish farms, gardens, greenhouses.

c) Artificial, as pest/unwanted species are reduced or eliminated.

Q2 a) $250 \times 180 = 45\ 000$. $45\ 000 \times 11 =$ **495 000 plants**

b) i) $(11 + 9 + 8 + 9 + 7) \div 5 =$ **8.8 plants**

ii) $45\ 000 \times 8.8 =$ **396 000**

iii) Lisa's. A larger sample gives a more accurate estimate.

Q3 The Manx cat and the Siamese cat are variations of the same species as they produce fertile offspring, but the lion and tiger are different species.

Page 25 — Populations and Competition

Q1 species, habitat, foods, resources, adapted

Q2 a) When the deer population increases, the wolf population increases shortly afterwards. The deer population then falls, and the wolf population then also decreases.

b) Because the wolves prey on the deer.

c) At point C

d) The deer population increased as there were fewer wolves to kill the deer.

Q3 a) In a parasitic relationship, parasites live off a host but give nothing back. In mutualism, both organisms benefit.

b) Any one of, e.g. Nitrogen-fixing bacteria in the root nodules of legumes. / 'Cleaner' animals, like the oxpecker birds found on the backs of buffalo.

Pages 26-27 — Evolution

Q1 a) A theory / hypothesis

b) E.g. It was so long ago that it's now very difficult to find any conclusive evidence for the theory or against it.

c) E.g. life began in a primordial swamp. Simple organic molecules joined to make more complex ones which eventually joined to give life forms.

Q2 a) minerals, decaying, skeletons, soft tissue, habitat, food

b) E.g. Because the conditions needed to make a fossil are very rare, so very few organisms become fossils after they die — most just decay away completely.

Q3 a) Scientists could claim that the skull came from a species whose ancestors were like chimps and which eventually evolved into humans — a 'missing link' to support the theory that humans evolved from chimp-like ancestors.

Section Four — Genes and Variation

b) Creationists would say that the skull belonged to a species of organism created individually by God, which is now extinct.

Q4 a) Rats that are resistant to warfarin are more likely to survive, so this genetic resistance to warfarin has been passed on from parent to offspring.

b) Darker tree trunks in polluted industrial areas would camouflage darker moths, so they have become more common in industrial areas where a dark colour is an advantage.

c) Some of the malaria organisms are resistant to the existing drugs, so they survive to reproduce and pass on the resistance genes to the offspring. A whole new population are soon resistant.

Q5 a) Extinct species are those that once lived but that don't exist any more.

b) We mainly know about extinct animals because we have found fossils of them. Also accept: We know more about some animals like mammoths because early people drew pictures of them, or about dodos because people wrote descriptions of them.

c) The environment changes too quickly (e.g. destruction of habitat). A new predator or disease kills them all (e.g. humans hunting them). They cannot compete with another species for food.

Page 28 — Natural Selection

Q1 All organisms face a struggle to survive.
The best adapted animals and plants are most likely to survive.
Some characteristics are passed on through reproduction from parent to offspring.

Q2
1. All giraffes had short necks.
2. A giraffe was born with a longer neck than normal. The long-necked giraffe was able to reach more food.
3. The giraffes competed for food from low branches. This food started to become scarce. Many giraffes died before they could breed.
4. The long-necked giraffe survived to have lots of offspring that all had longer necks.
5. More long-necked giraffes survived to breed, so more giraffes were born with long necks.
6. All giraffes had long necks.

Q3 People with the sickle cell gene are more likely to survive to reproduce in areas where large numbers of people are dying of malaria.

Q4 A

Section Four — Genes and Variation

Page 29 — Variation in Plants and Animals

Q1 a) genes
b) genes
c) both
d) both
e) both

Q2 a) No. Identical twins have exactly the same genes. Features like hair colour are controlled by genes, so you would expect the girls to have the same hair colour.

b) The difference in weight must be due to environment (e.g. eating more or exercising less), because the twins have exactly the same genes.

c) No. Identical twins have exactly the same genes, so if Anna had a birthmark then Zoe should too if it was genetic.

Q3 a) Environmental factors.
b) E.g. light intensity / wind exposure.
c) genetically
d) The difference may be caused by genetic factors or environmental factors, (e.g. water availability, soil fertility, amount of sunlight), or a combination of both.

Page 30 — Genes and Chromosomes

Q1 nucleus, chromosomes, DNA, genes
Q2 gene, chromosome, nucleus, cell
Q3 'Alleles' are different forms of the same gene.
Q4 There are two chromosome 7s in a human nucleus, one from each parent.
Q5 a) False
b) True
c) False

Page 31 — Sexual Reproduction and Variation

Q1 a) two
b) gametes
c) different
d) half as many

Q2 Sexual reproduction involves the production of gametes by each parent. Each gamete has half the normal number of chromosomes. The gametes fuse together and a baby with a full set of chromosomes is produced.

Q3 a) A change in the sequence of the DNA bases / a change in the genetic material of an organism.
b) E.g. nuclear radiation, X-rays, ultraviolet light, mutagenic chemicals.
c) carcinogens
d) They can stop proteins being produced, cause the wrong proteins to be produced or lead to the proteins being malformed.
e) Any two of, e.g. A mutation in a reproductive cell or an early stage embryo may cause the embryo to die. / The mutated cells might become cancerous. / A mutation could cause a genetic disorder.
f) A mutation may make the organism more suited to its environment and so make it more likely to survive.

Section Four — Genes and Variation

Page 32 — Genetic Diagrams

Q1 a) no
b) The dominant allele is shown by a capital letter (e.g. R) and the recessive allele by a lowercase letter (e.g. r).
c) homozygous
Q2 a) i) red eyes
ii) white eyes
iii) red eyes
iv) white eyes
b) i)

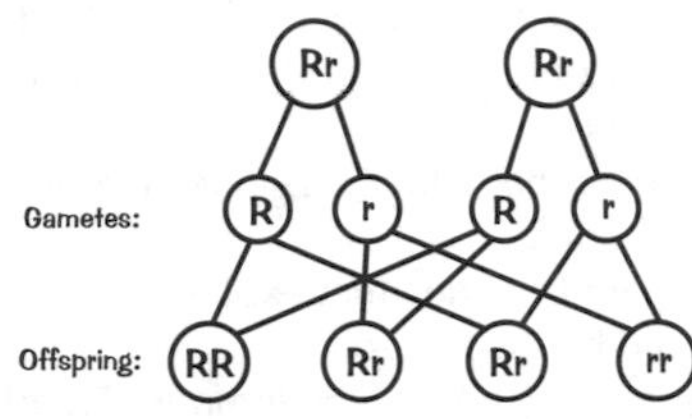

ii) 0.25 / 25% / 1 in 4 / ¼
iii) $0.75 \times 96 =$ **72**

Page 33 — Genetic Disorders

Q1 genetic, parents, allele, two, pancreas, carrier
Q2 a)

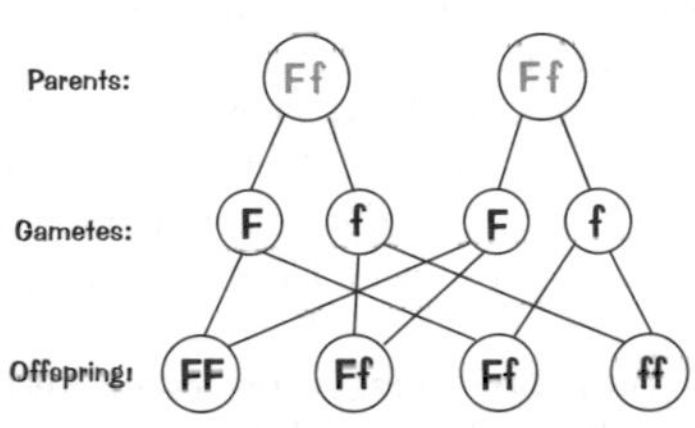

b) 0.25 / 25%
Q3 a) E.g. It might be very traumatic for the parents to have the baby knowing it is likely to die very young. / Pregnancy and childbirth can be risky for the mother's health and the risk may not be worth taking if the baby is likely to die. / If the baby does have the genetic disorder and is very ill, it may have a very poor quality of life.
b) E.g. Some people think it is wrong to take a life. / There is a small chance that the baby might be healthy. / An ill baby should have as much right to life as a healthy baby — it isn't moral to terminate the pregnancy because the child is likely to be ill.
Q4 a) gene therapy
b) No, this method only involves changing the genes in body cells, not reproductive cells. So the faulty gene would still be passed on to children.

Pages 34-35 — Cloning

Q1 a) clones
b) asexual
c) chromosome
d) E.g. bacterium, potato, strawberry plant, daffodil, spider plant/chlorophytum
Q2 a) The new skin cells came from the existing skin cells around the cut dividing to give new cells.
b) The new cells will be clones of the surrounding skin cells — because they are genetically identical they will look identical.
c) It took time for the cells to divide enough times to cover the cut completely.
Q3 a) Tissue culture and taking cuttings.
b) Advantages: e.g. identical copies of the best plants can be produced. Both methods quickly give new individuals.
Disadvantage: e.g. the gene pool is reduced.
Q4 a) embryo transplants
b) Sperm from the prize bull is used to artificially fertilise an egg from the prize cow. The resulting embryo is split at an early stage in its development to give lots of clones. These can then be implanted into the other cows in the herd to grow into calves.
c) The new calves will all be genetically identical to one another.
d) A reduced gene pool, meaning all the new calves will be vulnerable to the same diseases, etc.
Q5 a) False
b) False
c) True
d) False
e) True
Q6 E.g. some people think that using embryonic stem cells to treat disease is wrong, as the embryos are destroyed in the process. Other people think that helping humans who are already alive and suffering is more important than the potential lives of these embryos.

Page 36 — Genetic Engineering

Q1 gene, enzymes, enzymes, gene
Q2 a) The gene for human insulin production can be inserted into bacteria, which reproduce very quickly and produce a lot of insulin.
b) E.g. genetically engineering a crop so that it grows better in the developing nation, providing people with sufficient food. / Genetically engineering a commonly eaten food so that it contains a nutrient that people are lacking.
Q3 a) E.g. some people think that it's unnatural, and could cause unforeseen problems that could then be passed to future generations.
b) E.g. yes — because it can help people (e.g. it can help people with diseases like diabetes, it can help to grow more food in developing countries, etc.)
OR: No — there could be consequences that nobody has thought of yet / it might not be safe / it's not fully understood what the long-term effects might be.
Q4 The GM salmon might breed with wild salmon and the gene would spread through the wild population. Faster-growing salmon could disrupt natural food chains.

Section Five — Humans and the Environment

Section Five — Humans and the Environment

Page 37 — There's Too Many People

Q1 a)

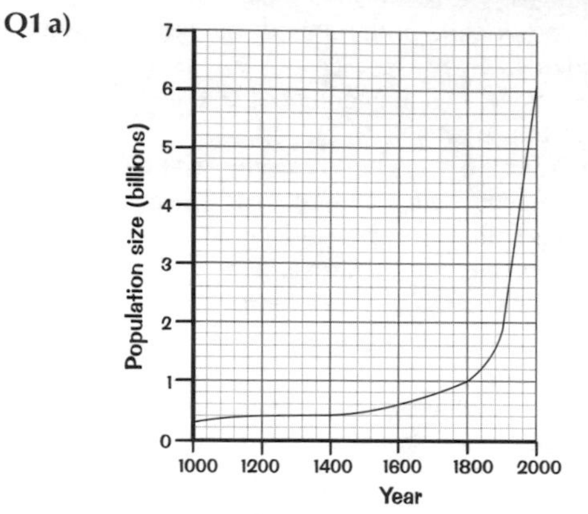

b) Any two of, e.g. Improvements in medicine meant that fewer people died of diseases. / More efficient farming methods produced more food so fewer people died of hunger. / Living standards improved over much of the world. / Hygiene and sanitation improved over much of the world.

Q2 a) i) John is more likely to live in the UK and Derek in Kenya.

ii) John buys more belongings, which use more raw materials to manufacture.
John has central heating in his home but Derek has a wood fire.
John drives a car and Derek rides a bicycle.

b) Any one of, e.g. use his car less / use his central heating less / recycle more waste / buy fewer new things.

Page 38 — The Greenhouse Effect

Q1 The greenhouse effect is needed for life on Earth as we know it. Increasing amounts of greenhouse gases may lead to global warming.

Q2 a) The greenhouses gases absorb much of the heat that is radiated away from Earth and re-radiate it in all directions. This keeps the atmosphere relatively warm.

b) All the energy radiated out by the Earth would be lost and the Earth would be much colder.

c) More greenhouse gas in the atmosphere means that more of the Sun's heat is trapped rather than radiated back out into space. This means the Earth gets warmer.

Q3 a) During deforestation, trees are often burnt to clear the land or as fuel, releasing carbon dioxide into the air. Microorganisms feed on the bits of dead wood that remain and release carbon dioxide when they respire. At the same time there is less carbon dioxide absorbed from the atmosphere. Trees take in carbon dioxide for photosynthesis, so if there are fewer trees there is less carbon dioxide absorbed.

b) Any two of, e.g. driving cars / industrial processes / generating electricity / cooking.

c) Gas: methane
Sources: rice-growing and cattle-rearing.

Page 39 — Climate Change

Q1 E.g. no, because no one knows for sure what will happen. Melting ice could disrupt ocean currents and the UK might get colder without the warm currents we have now.

Q2 Higher temperatures make ice melt.
Cold fresh water enters the ocean.
Ocean currents are disrupted.
Some areas (maybe the UK) get colder.

Q3 They might be right, but we can't know for sure from such small amounts of evidence. Both would need to carry out long-term studies using a lot more data.

Q4 a) **evidence:** data that either supports or contradicts a particular theory.
theory: a scientific idea that explains the facts but which could be disproved.

b) E.g. snow and ice cover / the temperature of the sea surface / the speed and the direction of ocean currents / atmospheric temperatures.

Page 40 — Air Pollution

Q1 a) The UV rays that enter through the hole kill the plankton that the whales feed on.

b) CFC gases (found in polystyrene, aerosols, air-conditioning units, fridges etc.)

c) UV rays increase the risk of skin cancer.

Q2 a) Carbon monoxide is formed when there is not enough oxygen for a hydrocarbon fuel to burn completely, which is more likely in an enclosed space.

b) Carbon monoxide is dangerous because it is poisonous. (It combines with haemoglobin preventing your blood from carrying as much oxygen.)

Q3 the greenhouse effect, sulfur dioxide, sulfuric, nitrogen oxides, nitric

Page 41 — Sustainable Development

Q1 a) Any two of, e.g. polluting the environment / overusing resources / reducing biodiversity / destroying habitats.

b) E.g. It can be difficult to clean areas that have been polluted. Some resources are non-renewable, so once they're used up you can never get them back. When species become extinct they are lost forever. Sensitive and ancient habitats like rainforest will not easily grow back.

Q2 a) Decreasing the number of different species in an area.

b) We could miss out on things like new medicines, foods or fibres. Ecosystems could become unbalanced, and all organisms as well as the environment itself could be badly affected.

Q3 a) An indicator species.

b) Any one of, e.g. collect samples of the same size / in the same way / at the same time of day.

c) Mayfly larvae prefer clean water and sludge worms can tolerate water that contains sewage.

d) E.g. sewage is full of bacteria, which use up a lot of oxygen. Animals like mayfly larvae might not have enough oxygen to survive.

Page 42 — Conservation and Recycling

Q1 a) True
b) True
c) False
d) False
e) False

Q2 coppicing — cutting trees down to just above ground level
reforestation — replanting trees that have been cut down in the past
replacement planting — new trees are replanted at the same rate that others are cut down

Q3 a) Recycling is when materials are reprocessed to make new goods instead of being thrown away.
b) Any three of, e.g. glass / paper / plastics / metal (or example, e.g. aluminium).
c) less energy

Q4 E.g. conservation measures may help protect our food supply for the future, e.g. fishing quotas. Protecting habitats may help prevent useful species (e.g. medicinal plants) from going extinct. Recycling metals and paper means using less raw materials, cuts energy use and reduces pollution.

Section Six — Cells and Cell Functions

Page 43 — Cells

Q1 a) True
b) False
c) True
d) False

Q2 a) Plant, animal
b) cell wall
c) Both plant and animal cells
d) photosynthesis, glucose

Q3 Lots of chloroplasts — for photosynthesis
Tall shape — gives a large surface area for absorbing CO_2
Thin shape — means you can pack more cells in at the top of the leaf

Q4 stomata, turgid, photosynthesis, flaccid

Q5 a) The **nucleus** contains genetic material/chromosomes/genes/DNA. Its function is controlling the cell's activities.
b) **Chloroplasts** contain chlorophyll. Their function is to make food by photosynthesis.
c) The **cell wall** is made of cellulose. Its function is to support the cell and strengthen it.

Page 44 — DNA

Q1 nucleus, helix, nucleotides, base, adenine/guanine, adenine/guanine

Q2
1. The DNA double helix 'unzips' to form two single strands.
2. Free-floating nucleotides join on where the bases fit.
3. Cross links form between the bases of the nucleotides and the old DNA strands.
4. The new nucleotides are joined together.
5. The result is two molecules of DNA identical to the original molecule of DNA.

Q3

A	C	T	G	C	A	A	T	G
T	G	A	C	G	T	T	A	C

Q4
1. Collect the sample for DNA testing.
2. Cut the DNA into small sections.
3. Separate the sections of DNA.
4. Compare the unique patterns of DNA.

Page 45 — Making Proteins

Q1 a) amino acids
b) liver
c) ribosomes
d) three

Q2
1. The DNA strand unzips.
2. A molecule of RNA is made using DNA as a template.
3. RNA moves out of the nucleus.
4. RNA joins with a ribosome.
5. Amino acids are joined together to make a polypeptide.

Q3 a) DNA is too big to move out of the nucleus into the cytoplasm. RNA is small enough.
b) E.g. RNA is smaller / RNA is only one strand / RNA contains the base uracil instead of thymine.
c) a gene
d) The number and order of amino acids gives a protein a particular shape and therefore a particular function.
e) Our bodies can change some amino acids into others — this is called transamination.
f) DNA controls which genes are switched on or off. This determines which proteins the cell produces (e.g. haemoglobin), which determines what type of cell it is (e.g. a red blood cell).
g) A sequence of three bases in a strand of DNA codes for a particular amino acid. Amino acids are stuck together to make proteins, following the order of the code on the DNA.

Section Seven — Organs and Systems 1

Page 46 — Enzymes

Q1 a) An enzyme is a biological catalyst. It increases the speed of a reaction without being changed or used up in the reaction.

b)

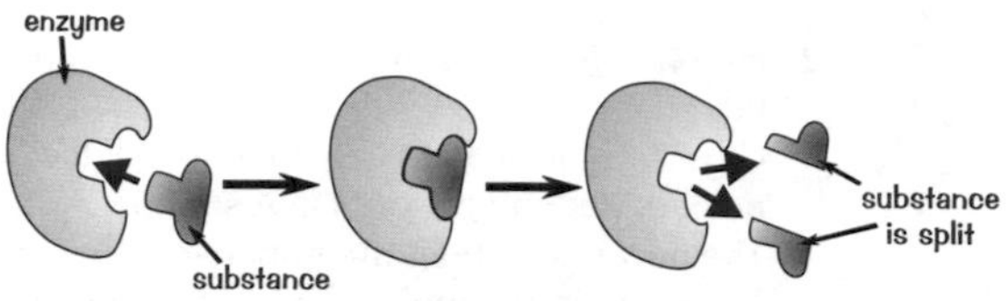

Q2 a) False
b) True
c) True
d) False

Q3 a)

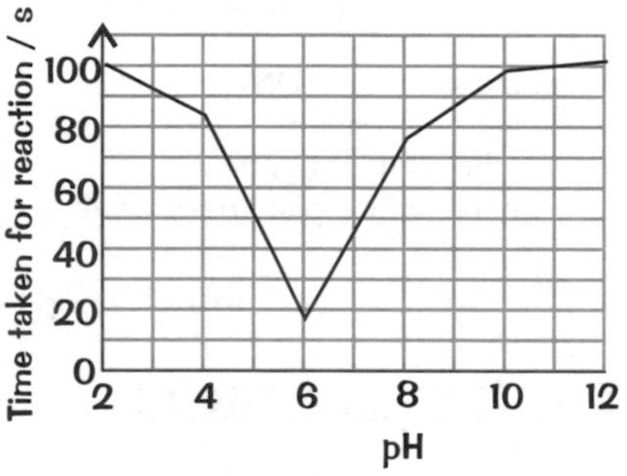

b) About pH 6.
c) At very high and very low pH levels the bonds in the enzymes are broken/the enzyme is denatured, meaning that it is damaged so it can't speed up the reaction.
d) No. This enzyme works very slowly at pH 2/ in strongly acidic conditions.

Page 47 — Diffusion

Q1 random, higher, lower, bigger, large, less
Q2 a) False
b) True
c) True
d) False

Q3 a)

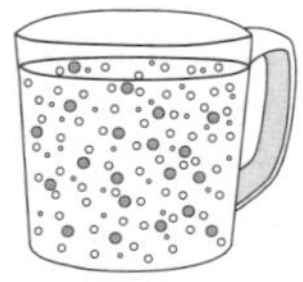

(The sugar particles will have spread out evenly.)

b) i) The rate of diffusion would **be faster**.
ii) The rate of diffusion would **be faster**.
iii) The rate of diffusion would **be slower**.
c) The sugar particles will move from an area of higher concentration (the sugar cube) to an area of lower concentration (the tea).

Page 48 — Diffusion in Cells

Q1 amino acids, small, diffusion, lower, out of
Q2 nerve, synapse, impulse, transmitter, diffuses, binds, receptor
Q3 a) and **b)**

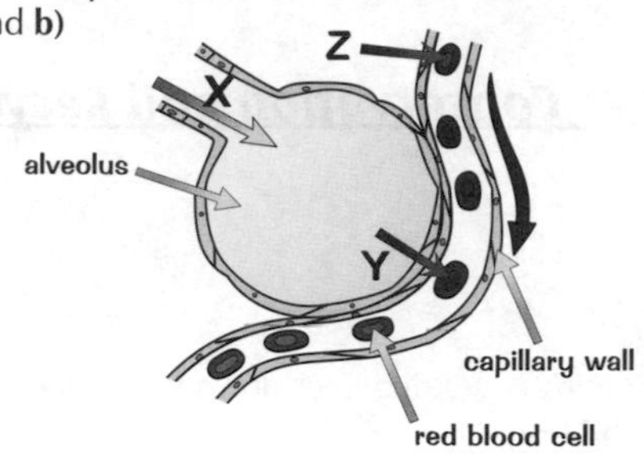

c) Into the blood. Although the person is breathing out, the concentration gradient still causes net diffusion from the alveolus into the bloodstream.

Page 49 — Osmosis

Q1 a) A membrane that only allows certain substances to diffuse through it.
b) Side B
c) From B to A
d) The liquid level on side B will **fall**, because water will flow from side B to side A by osmosis.
Q2 a) Tissue fluid.
b) More water will move into the cells as the tissue fluid has a higher water concentration than the cells.
c) After a while there's an equal concentration of water molecules on either side of the cell membranes, so the molecules flow in both directions at the same rate and there's no overall change.

Section Seven — Organs and Systems 1

Page 50 — Respiration and Exercise

Q1 a) True
b) False
c) True
d) False
e) True
f) True
g) False
Q2 a) glucose + oxygen → carbon dioxide + water (+ energy)
b) glucose → lactic acid (+ energy)
Q3 a) 45 – 15 = **30** breaths per minute
b) During exercise, aerobic respiration in the muscles increases to provide more energy. The breathing rate increases to provide more oxygen for respiration in the muscles.

Section Seven — Organs and Systems 1

c) Because Jim has an oxygen debt after the race. Extra oxygen is needed to break down the lactic acid produced by anaerobic respiration in his muscles during the race.

Q4 a) Any two of, e.g. large surface area / moist surface / thin walls / permeable walls / good blood supply.

b) It would decrease, because the oxygen concentration in the alveoli would fall (so there would be a smaller concentration gradient).

Page 51 — Enzymes and Digestion

Q1

a) protein $\xrightarrow{\text{protease}}$ amino acids

b) fat $\xrightarrow{\text{lipase}}$ glycerol + fatty acids

c) carbohydrate $\xrightarrow{\text{amylase}}$ simple sugars

Q2

Amylase	Protease	Lipase	Bile
salivary glands pancreas small intestine	stomach pancreas small intestine	pancreas small intestine	liver

Q3 a) gall bladder, small intestine, neutralises, enzymes, fat

b) Emulsification breaks fat into smaller droplets, which gives a larger surface area for lipase to work on, speeding up digestion.

Page 52 — The Digestive System

Q1

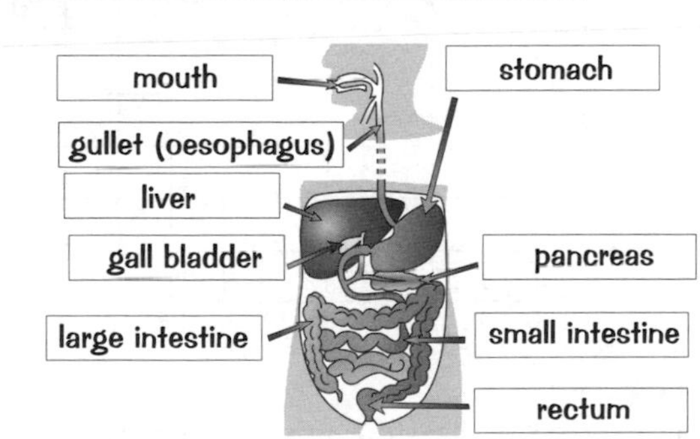

Q2 a) Stores bile / releases bile into the small intestine.

b) Produces enzymes (protease, amylase and lipase) / releases enzymes into the small intestine.

c) Produces bile, which emulsifies fats and neutralises stomach acid.

d) Absorbs excess water from food.

Q3 a) i) capillary network

ii) longitudinal muscle

iii) villus

iv) circular muscle

b) i) Provides a very large surface area for diffusion to happen across.

ii) To allow time for all the food to be digested and absorbed before it reaches the end. It also increases the surface area.

Page 53 — Functions of the Blood

Q1 a) False

b) False

c) True

d) False

e) True

Q2 a)

Substance	Travelling from	Travelling to
Urea	**liver**	**kidneys**
Carbon dioxide	**body cells**	**lungs**
Glucose	**gut**	**body cells**

b) Any six of, e.g. red blood cells / white blood cells / platelets / water / amino acids / hormones / antibodies / antitoxins

Q3 large, nucleus, haemoglobin, oxygen, oxyhaemoglobin, flexible (accept small)

Q4 a) antibodies

b) antitoxins

c) White blood cells have a flexible shape which enables them to wrap around the microorganism, before digesting it using enzymes.

Page 54 — Blood Vessels

Q1 artery — vessel that takes blood away from the heart
capillary — microscopic blood vessel
lumen — hole in the middle of a tube
vein — vessel that takes blood towards the heart

Q2 a) veins

b) capillaries

c) arteries, veins

Q3 a) E.g. for making cell membranes.

b) saturated fat (accept fat)

c) Cholesterol can build up in your arteries as plaques. These restrict and may block the lumen, preventing blood getting through. This can lead to a heart attack or stroke.

Q4 a) Any two of, e.g. the vein has a wider lumen / thinner wall / valves.

b) The vein

c) Length of blood vessel, because it's the dependent variable.

d) To make his experiment a fair test (the vessels are the same age, from the same animal, etc.)

Page 55 — The Heart

Q1 a) pulmonary artery

b) vena cava

c) right atrium

d) tricuspid valve

e) right ventricle

f) aorta

g) pulmonary vein
h) semi-lunar valve
i) bicuspid valve
j) left ventricle

Q2 a) False
b) True
c) True
d) True

Q3 a) Mammals have two separate circulation systems — one to the lungs and one to the rest of the body.
b) The atria only have to pump blood into the ventricles, whereas the ventricles have to pump blood out of the heart at a very high pressure, so they need to have thicker muscular walls to do this.
c) The right ventricle only has to pump blood to the lungs, so it doesn't need as much muscle. The left ventricle has to pump blood to the rest of the body, so it needs to be more muscular.

Q4 They could give the patient drugs that suppress the patient's immune response.

Q5 pacemaker, irregular, pacemaker, valves

Page 56 — The Kidneys and Homeostasis

Q1 a) False
b) True
c) True
d) False
e) True

Q2 a) They're taken into the body in food and drink, and then absorbed into the blood.
b) E.g. too much or too little water would be drawn into the cells, which would damage the cells. / Nerves and muscles wouldn't work properly.
c) In the sweat.

Q3 a) Any three of, e.g. in sweat / in our urine / in the air we breathe out / in faeces.
b)

	Do you sweat **a lot** or **a little**?	Is the amount of urine you produce **high** or **low**?	Is the urine you produce **more** or **less** concentrated?
Hot Day	A lot	low	more
Cold Day	A little	high	less

c) Sheona would have lost a lot of water as sweat when she got hot during the run. To try to avoid dehydration her kidneys will have excreted as little water as possible, meaning that there was little urine and it was concentrated.

Page 57 — The Pancreas and Diabetes

Q1 a) True
b) False
c) False
d) False

Q2 a) From the pancreases of other dogs.
b) i) The blood sugar levels dropped.
ii) insulin

Q3 a) Diabetics used to use insulin from animals such as pigs, which sometimes caused problems. Now they can use human insulin produced by genetically modified bacteria.
b) Diabetics used to use glass syringes that had to be boiled. Now they can use disposable syringes that are already sterile. There are also needle-free devices.

Q4 a) A pancreas transplant.
b) Risk of infection, risk of rejecting the organ, having to take immunosuppressive drugs.
c) Any two of, e.g. transplanting just the cells which produce insulin / artificial pancreases / using stem cells.

Section Eight — Growth and Development

Page 58 — Growth

Q1 a) Any two of, height, length, width, circumference
b) The weight of an organism including all the water in its body.
c) It varies a lot, depending on how much water the organism has gained or lost during the day.

Q2 Any two of, e.g. some types of worm are able to grow a new 'tail' if cut in half. / A young spider can regrow a leg. / Some reptiles can regrow a lost leg or tail.

Q3 E.g. animals tend to grow when they are young while plants grow continuously. Growth in animals happens by cell division, whereas in plants growth in height is mainly due to cell elongation.

Q4 a)

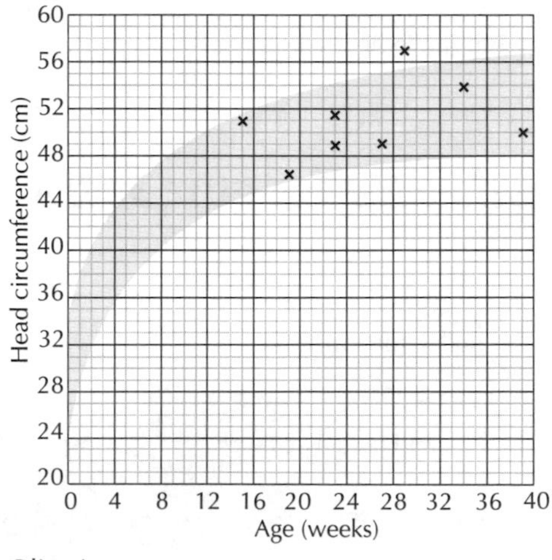

b) Oliver's
c) To determine whether the baby is growing too fast or too slowly, which could alert the doctor to developmental problems. / If the baby's head is very large or small relative to the rest of its body it could alert the doctor to developmental problems.

Page 59 — Cell Division — Mitosis

Q1 a) True
b) False
c) False
d) True
e) True
f) True

Section Eight — Growth and Development

g) True
Q2 a) DNA is spread out in long strings.
b) DNA is duplicated and forms X-shaped chromosomes.
c) The chromosomes line up at the centre of the cell and cell fibres pull them apart. The left and right arms go to opposite ends of the cell.
d) Membranes form around each of the sets of chromosomes. They become the nuclei of the two new cells.
e) The cytoplasm divides to form two new cells.
Q3 a) The Hayflick limit is the limit to the number of times a cell can divide.
b) Stem cells and cancer cells.

Page 60 — Cell Division — Meiosis

Q1 a) Meiosis
b) Mitosis, Meiosis
c) Meiosis
d) Meiosis
Q2 a) The DNA is spread out in long strands. Before the cell starts to divide it duplicates its DNA to produce an exact copy.
b) For the first meiotic division the chromosomes line up in their pairs across the centre of the cell.
c) The pairs are pulled apart, mixing up the mother and father's chromosomes into the new cells. This creates genetic variation.
d) The chromosomes line up across the centre of the nucleus ready for the second division, and the left and right arms are pulled apart.
e) There are now 4 gametes, each containing half the original number of chromosomes.
Q3 acrosome containing enzymes — to digest the membrane of the egg cell
produced in large numbers — to give many chances for fertilisation
small with long tails — so they can swim to the egg
lots of mitochondria — to provide the energy needed to move.

Page 61 — Sexual Reproduction — Ethics

Q1 a) 24 weeks
b) It was based on the point that a foetus became able to survive outside the womb (sometimes only with medical help).
c) When the pregnancy is putting the mother's health at serious risk, or if there is a major foetal abnormality.
d) E.g. some people think that human life begins at fertilisation so ending a pregnancy at any time would be the same as killing a human being.
e) E.g. some people believe that a foetus can feel pain at 7 weeks old, so termination should not be allowed after this time. / Some people believe the foetus can't feel pain until the pain receptors are connected to the brain at about 26 weeks, so termination should be allowed up to this time. / With medical advances foetuses are becoming viable earlier in the pregnancy so some people feel the limit should be dropped to match this.

Q2 a) E.g. embryonic screening implies that people with genetic disorders are 'undesirable' and this may increase prejudice. / The rejected embryos are destroyed. Each one is a potential human life. / There's a risk that embryonic screening could be taken too far, e.g. parents might want to choose embryos who fulfil their vision of the ideal child.
b) E.g. if embryonic screening means healthy children are born, then this stops the suffering associated with many genetic disorders. / During IVF, most of the embryos are destroyed anyway — screening just allows the selected one to be healthy.

Page 62 — Stem Cells and Differentiation

Q1 a) Cells that are adapted to perform a particular job.
b) The process by which a cell changes to become specialised.
c) Cells that aren't yet specialised and can develop into different types of cell depending on what instructions they get.
Q2 Embryonic stem cells can differentiate into any type of body cell. Adult stem cells are less versatile — they can only turn into certain types of cell.
Q3 E.g. people with some blood diseases (e.g. sickle cell anaemia) can be treated using bone marrow transplants. Bone marrow contains stem cells that can turn into new blood cells to replace the faulty ones.
Q4 diabetes — insulin-producing cells
paralysis — nerve cells
heart disease — heart muscle cells
Q5 a) E.g. stem cell research may lead to cures for a wide variety of diseases. / The embryos used are usually unwanted ones that would be destroyed anyway.
b) E.g. embryos shouldn't be used for experiments as each one is a potential human life.

Page 63 — Growth in Plants

Q1 a) False
b) True
c) False
d) True
Q2 a) auxins
b) At the tips of shoots and roots.
c) i) In the shoot auxin moves away from the light and stimulates growth on the shaded side, making the shoot bend towards the light.
ii) In the root auxin moves to the lower side of the root and inhibits growth, making the root grow downwards.
Q3 a) The flowers aren't pollinated. Instead a growth hormone is applied.
b) E.g. unripe fruit is firmer, so it's less easily damaged and bruised during picking and transport.
c) They are sprayed with a ripening hormone during transport.

Section Nine — Plants and Energy Flow

Page 64 — Selective Breeding

Q1 Fruit plant: any two of, e.g. larger fruit / greater yield / sweeter/better tasting fruit / fruit with better colour / fruit that ripens more quickly / plant more resistant to disease / faster growth rate.
Ornamental house plant: any two of, e.g. more colourful leaves/flowers / more scented flowers / more resistant to disease / faster growth rate.

Q2 E.g. there's less variety in the gene pool of the organism — this means that they will have similar levels of disease resistance and some diseases may be able to wipe out the whole lot. / The characteristics selected for by humans may not be beneficial for the organism, e.g. mastitis in cows due to greater milk production.

Q3 a) Female sheep that produce large numbers of offspring could be bred with rams with mothers that produced large numbers of offspring.

b) Normal wheat plants with a good grain yield could be bred with dwarf wheat plants (that are more able to resist the wind and rain).

Q4 a) Yes, the average milk yield has increased over the generations.

b) 5375 – 5000 = **375 litres**

Page 65 — Adult Cloning

Q1 a) removing and discarding a nucleus = **A**
implantation in a surrogate mother = **D**
useful nucleus extracted = **B**
formation of a diploid cell = **C**

b) mitosis

c) E.g. the embryo may not develop normally. /
The clone may be unhealthy and die prematurely.

Q2 a) More animals that can produce the blood clotting agent would increase the yield of the agent, and the process of genetic engineering would only have to happen once.

b) Some people might think it is unethical to clone an animal. / Cloning is a new science and we might be unaware of possible problems associated with it.

Q3 a) Transplanting organs from animals into humans.

b) Animals could be genetically engineered to make organs that can be safely transplanted into humans, and then these animals could be cloned to increase the number of organs available for transplantation.

Q4 E.g. lots of women will have to be willing to donate eggs. There would have to be a lot of surrogate pregnancies. There could be high rates of miscarriage and stillbirth. At the moment, evidence suggests that animal clones are sometimes unhealthy and die prematurely — it could be the same for humans. We are playing about with genetics that we don't completely understand. The clone may be psychologically damaged knowing that it's a clone of another human being.

Section Nine — Plants and Energy Flow

Page 66 — Photosynthesis

Q1 a) carbon dioxide + water → glucose + oxygen

b) chloroplast — the structure in a cell where photosynthesis occurs
chlorophyll — a green pigment needed for photosynthesis
sunlight — supplies the energy for photosynthesis
glucose — the food that is produced by photosynthesis

Q2 leaves, energy, convert, cells, fructose, sucrose, fruits, cellulose, walls, lipids

Q3 a) 00.00 (midnight)

b) There's no light at night so photosynthesis can't happen.

c) Plants use the food they have stored as starch.

d)

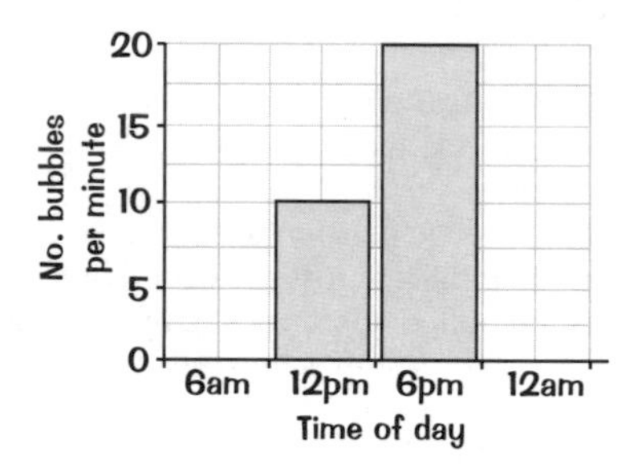

Page 67 — Rate of Photosynthesis

Q1 a) E.g. light intensity, CO_2 concentration, temperature

b) A factor that stops photosynthesis from happening any faster.

c) E.g. time of day (such as night time) / position of plant (such as in the shade).

Q2 a)

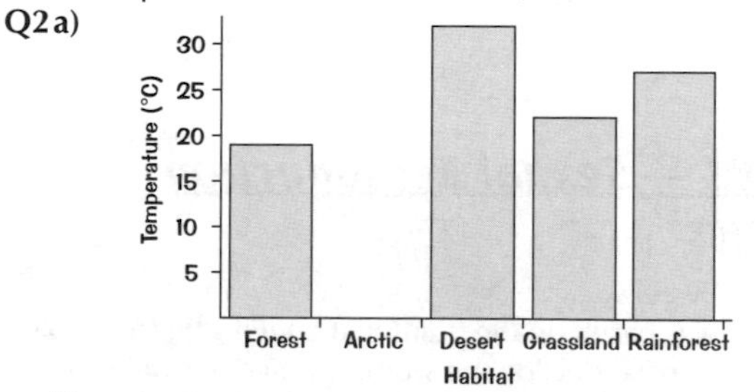

b) In the Arctic.

c) The temperatures are low there, so the rate of photosynthesis will be slow because the enzymes needed for photosynthesis will be working slowly.

Q3 a) Increasing the concentration of CO_2 increases the rate of photosynthesis up to a certain point.

b) The rate of photosynthesis doesn't continue to increase because something else (e.g. temperature) is the limiting factor.

Section Nine — Plants and Energy Flow

Page 68 — Leaf Structure

Q1 A — palisade mesophyll layer
B — upper epidermis
C — waxy cuticle
D — stoma
E — guard cell

Q2 a) photosynthesis
b) respiration

Q3 a) A — oxygen / water vapour
B — water vapour / oxygen
C — carbon dioxide
b) The leaf would not be photosynthesising as there's no light. This means that mainly carbon dioxide would be diffusing out of the leaf and oxygen diffusing in. Less water will diffuse out.

Q4 a) Provide a large surface area for gas exchange. / Lets gases easily move between cells.
b) Means a large surface area is exposed to light.
c) Deliver water (and nutrients) to leaf cells.
d) Contain chlorophyll to absorb light energy.

Page 69 — Transpiration

Q1 a) b) c)

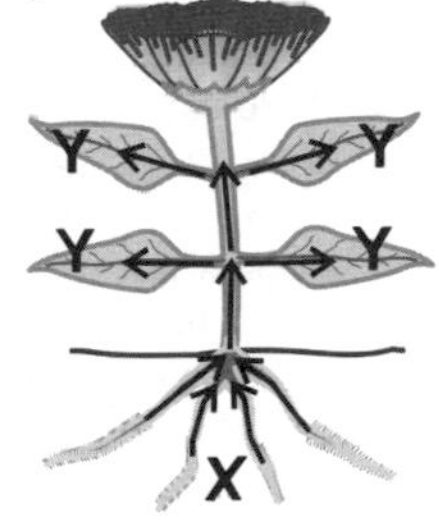

Q2 a) False
b) True
c) True
d) False

Q3 leaves, diffusion/evaporation, evaporation/diffusion, leaves, xylem, roots, transpiration

Q4 Any three of, e.g. it helps to keep the plant cool / the plant has a constant supply of water for photosynthesis / it creates a turgor pressure which provides support / essential minerals from the soil are constantly brought into the plant with the water.

Q5 a) Open, because the plant will be using the sunlight to photosynthesise and so it needs to take in CO_2.
b) Most of the stomata close at night, which avoids losing water by transpiration.
c) The plant won't lose even more water by transpiration, but it won't be able to photosynthesise either.

Page 70 — Water Flow in Plants

Q1 a) E.g. stomata are mostly found on the lower surface of the leaf where it's cool and dark. There is a waterproof waxy cuticle covering the leaf surface.
b) They have no stomata on the upper epidermis and fewer, smaller ones on the underside of the leaf/ spine.

Q2 a) A — normal
B — turgid
C — flaccid
D — plasmolysed
b) The cells lose water and their contents are no longer able to push up against the cell wall to support the cell (turgor pressure decreases).
c) The inelastic cell wall doesn't collapse completely.

Q3 a) minerals, water
b) support
c) It has very thick side walls that are strong and stiff.

Q4

XYLEM VESSELS	PHLOEM VESSELS
transport water	made of living cells
have no end-plates	have end-plates
made of dead cells	transport food

Page 71 — Minerals for Healthy Growth

Q1 magnesium — for making chlorophyll
potassium — for helping enzymes to function
phosphates — for making DNA and cell membranes
nitrates — for making proteins

Q2 a) root hair cell
b) Absorbing water and minerals from the soil.
c) It has a very large surface area for absorbing the minerals and water.
d) The soil generally has a lower concentration of minerals than the root hair cells. Diffusion only takes place from areas of higher concentration to areas of lower concentration.
e) The cells use active transport to absorb the minerals. (This requires energy, which is released by respiration.)

Q3 a) nitrates
b) Poultry manure — it contains the highest percentage of nitrogen.

Page 72 — Pyramids of Number and Biomass

Q1

Feature	Pyramid of numbers	Pyramid of biomass
Values for mass are shown at each level.		✓
Nearly always a pyramid shape.		✓
Each bar represents a step in a food chain.	✓	✓
Always starts with a producer.	✓	✓
Can only have 3 steps.		

Q2 a) A
b) C
c) The total mass of the organisms decreases at each trophic level as shown by this pyramid.

Q3 a)

ladybirds
aphids
oak tree

b) There is a single, large organism at the bottom of the food chain.

Section Nine — Plants and Energy Flow

Q4 a) The population of lions might increase as there would be more food for them.

b) The number of zebras might decrease as they would be competing with more gazelles for food. / The number of zebras might decrease as there are more lions to eat them. / The number of zebras might increase as there are more gazelles for the lions to eat instead.

Page 73 — Energy Transfer and Energy Flow

Q1 a) energy
b) Plants, photosynthesis
c) eat
d) respiration
e) lost, movement
f) inedible, hair

Q2 a) i) 100 000 – 90 000 = **10 000 kJ**
ii) (10 000 ÷ 100 000) × 100 = **10%**
b) i) (5 ÷ 100) × 1000 = **50 kJ**
ii) 1000 – 50 = **950 kJ**

Q3 a) (2070 ÷ 103 500) × 100 = **2%**
b) 2070 ÷ 10 = 207
207 – (90 + 100) = **17 kJ**
c) Any two of, e.g. heat loss / movement / excretion
d) So much energy is lost at each stage of a food chain that there's not enough left to support more organisms after about five stages.

Page 74 — Biomass and Fermentation

Q1 E.g. eat the biomass / feed it to livestock / grow the seeds of plants / use the biomass as fuel.

Q2 a) More trees can be planted to replace those that are burnt. The new trees are photosynthesising so remove CO_2 from the atmosphere.
b) A digester or fermenter.
c) It is burnt and used for heating or to generate electricity.

Q3 a) False
b) False
c) True
d) False
e) False

Q4 a) A protein from a fungus used to make meat substitutes.
b) Any two of, e.g. they grow very quickly / they're easy to look after / they can use waste products from agriculture and industry as food / food can be produced whatever the climate or terrain.
c) i) To keep the fermenter at the correct temperature.
ii) To supply oxygen for aerobic respiration of the fungi.
iii) To mix the contents / stop the microorganisms sinking to the bottom.

Page 75 — Managing Food Production

Q1 a) Wheat → Human
There are fewer steps in this food chain so less energy is lost (energy is lost at every stage in a food chain).

Q2 a) It creates a favourable environment for the spread of disease, e.g. avian flu. / The animals may be uncomfortable.
b) When the animals are eaten, the antibiotics will enter humans. This may help microbes that infect humans to develop resistance to the antibiotic.
c) This often means using power from fossil fuels (which wouldn't be used if the animals were grazing naturally.)

Q3 a) It's the breeding season — lots of small fish are born, so the average size decreases.
b) E.g. a high number of fish lice, a high water pH, predators may eat the fish, competition for food by other species.
c) In this experiment the fish farm (Ecosystem A) produces larger fish and more fish survive.

Page 76 — Pesticides and Biological Control

Q1 a) Pesticides are chemicals that are used to kill creatures to stop them damaging crops, gardens, buildings, etc.
b) Any one of, e.g. they may kill organisms that are not pests / they can disrupt food chains / they could be harmful to humans eating the food that has been sprayed.

Q2 The frog population would decrease as there's less for them to eat, so the foxes would have to eat more rabbits. The populations of both rabbits and foxes could therefore decrease too.

Q3 a) Biological control is when living organisms rather than chemicals are used to control a pest species.
b) Any two of, e.g. using a predator — ladybirds are introduced into gardens to keep the numbers of aphids down. / Using a parasite — certain types of wasps and flies produce larvae inside a host insect, which eventually kills it. / Using a disease — myxomatosis is a virus introduced to control the rabbit population.
c) Advantage: e.g. no chemicals are used / only the pest species is usually affected, so there's less disruption of food chains / there's less risk to people eating the food.
Disadvantage: e.g. it's slower to take effect / more difficult to manage / may not kill all the pests / there is a risk that control organisms could become pests themselves or drive out native species.

Q4 a) The birds of prey ate animals that had eaten the crops.
b) Each small animal ate a lot of crops, and each bird ate a lot of the small animals. If the chemical was not excreted it would build up through the food chain and reach toxic levels.

Section Ten — Organs and Systems 2

Page 77 — Alternatives to Intensive Farming

Q1 a) A mixture of fertilisers dissolved in water / a nutrient solution.
b) E.g. tomatoes / cucumbers
c) i) advantage
ii) disadvantage
iii) advantage
iv) disadvantage
v) disadvantage

Q2 a) **Insecticides**: Alternative — biological control / crop rotation / varying seed planting times.
Advantage: No chemicals used so safer for humans eating the crops / less likely to disrupt food chains / less likely to kill harmless or beneficial organisms.
b) **Herbicides**: Alternative — weeding.
Advantage: No chemicals used so safer for humans eating the crops / less likely to disrupt food chains / doesn't kill harmless or beneficial organisms.

Q3 a) Cost — the farmer won't have to pay for artificial fertilisers and herbicides. However, he or she might have to pay extra staff. The yields could be lower so there may not be as much profit. This may mean that more land is needed to get the same amount of crop, which is expensive.
b) Labour — weeding is much more labour intensive than simply spraying a field with a chemical. Spreading manure and compost could also take more work than spraying a liquid over the crop.
c) Environment — Without the use of herbicides there will be less disruption to food chains. A greater variety of wildlife will be able to survive on the farm.

Page 78-79 — Recycling Nutrients

Q1 a) b) c)

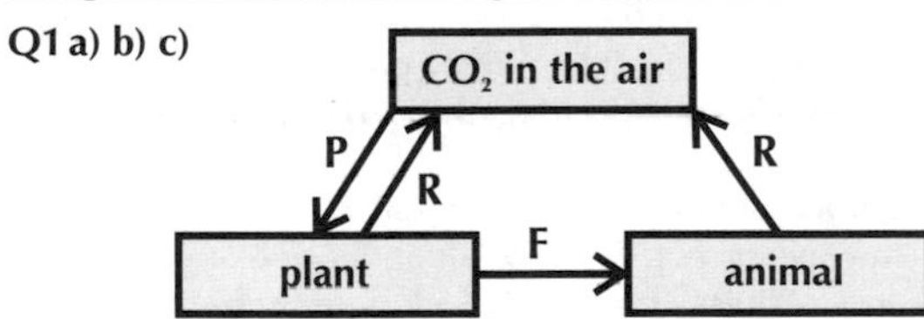

Q2
1. Plants take up minerals from the soil.
2. Plants use minerals and the products of photosynthesis to make complex nutrients.
3. Nutrients in plants are passed to animals by feeding and used in respiration to provide energy.
4. Plants and animals die.
5. Materials are recycled and returned to the soil by decay.

Q3 a) by photosynthesis
b) by animals feeding on other organisms
c) respiration

Q4

Type of organism involved in decay	Example	How they help in decay
Detritivores	**E.g. Earthworms / maggots / woodlice**	**They break up decaying material into smaller bits, which provides a larger surface area for smaller decomposers to work on.**
Saprophytes	Bacteria / fungi	**They secrete digestive enzymes onto decaying material which breaks down the material into bits that can be absorbed.**

Q5 a) protein
b) 78%
c) an unreactive gas

Q6 Plants — From nitrates in the soil
Animals — By eating other organisms
Bacteria in soil — By breaking down dead organisms and animal waste

Q7 a) Decompose proteins and urea into ammonia.
b) Turn ammonia in decaying matter into nitrates which plants can use.
c) Turn nitrates back into nitrogen gas.
d) Turn nitrogen gas into nitrogen compounds that plants can use.

Q8 a) Nitrogen-fixing bacteria converting nitrogen from the air into nitrogen compounds in the soil.
b) Nitrifying bacteria converting ammonia into nitrates.
c) Denitrifying bacteria converting nitrates in the soil into nitrogen gas.

Q9 Legume plants have root nodules that contain nitrogen-fixing bacteria. These bacteria convert nitrogen gas into nitrogen compounds, increasing the fertility of the soil.

Section Ten — Organs and Systems 2

Page 80 — Solute Exchange — Active Transport

Q1 a)

Feature	Diffusion	Osmosis	Active transport
Substances move from areas of higher concentration to areas of lower concentration	✓	✓	
Requires energy			✓

b) Diffusion can happen with **any** substance where the particles are free to move — osmosis is the movement of **water** particles only.

Q2 a) The meal had to be transported to the gut and digested to release the glucose.
b) The initial rate of absorption is higher for the meal contain 40 g of starch. This is shown by the higher initial gradient for this meal, on the graph.
c) The rate decreased because the concentration of glucose in the gut was decreasing.
d) Theo fasted before this investigation, so the concentration of glucose in his blood would have been low. After eating, the concentration of glucose would be high in the gut, so diffusion is likely to have been responsible.

Page 81 — The Respiratory System

Q1 a)

Section Ten — Organs and Systems 2

b) in, intercostal, diaphragm, flattens, up/out, out/up, increases, decreases, drawn into.
relax, ribcage, down/in, in/down, volume, decreases, increases, forced out of.

Q2 a)

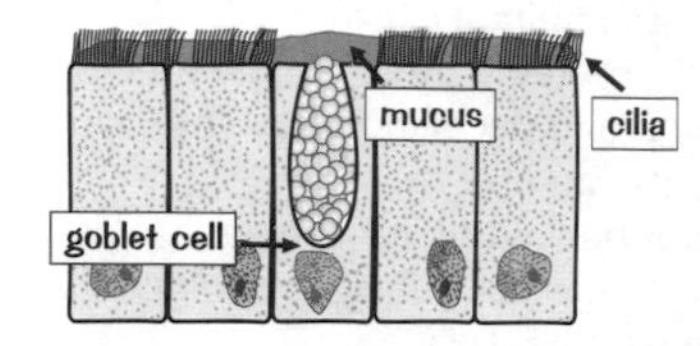

b) i) produce mucus
ii) Catches dirt and microbes before they reach the bottom of the lungs.
iii) Catch dirt and microbes and push mucus out of the lungs as phlegm.
c) The respiratory tract is a dead end so microbes cannot be flushed out.

Page 82 — Lung Capacity and Disease

Q1 a) a spirometer
b) a spirogram
c) tidal volume
d) Total lung capacity minus residual volume.
e)
f) E.g.

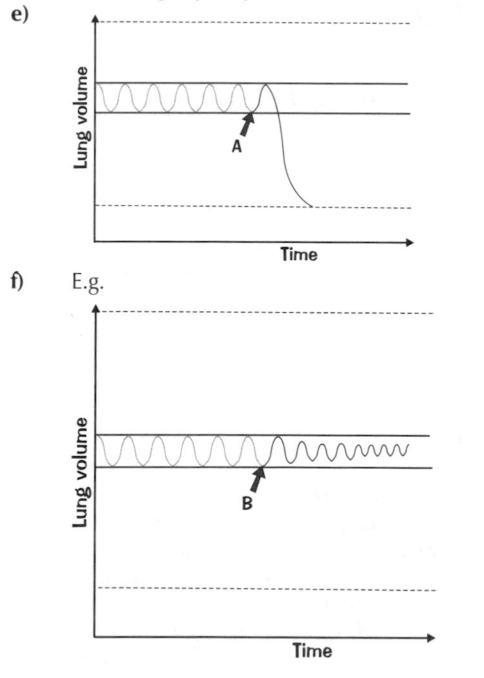

Q2 a) Cystic fibrosis is inherited/genetic. A thick mucus is produced which can block the lungs. This makes breathing difficult and can make sufferers more prone to infection.
b) Asbestosis is caused by breathing in asbestos, usually in an industrial setting. The lung tissue becomes scarred, causing breathing problems.
c) Lung cancer can be caused by lifestyle, e.g. smoking. Cells divide out of control, forming tumours.
Q3 muscles, contract, difficult, wheezing/coughing/a tight chest, wheezing/coughing/a tight chest, wheezing/coughing/a tight chest, relaxants

Page 83 — The Circulation System

Q1 a) The heart has two sides, which each pump blood to different areas of the body.
b) i) right
ii) deoxygenated
c) i) left
ii) oxygenated
d) The left side has to create more pressure to force blood to travel to the organs of the entire body. The right side of the heart only has to pump blood to the lungs, which is a shorter distance.
Q2 A — Pulmonary artery
B — Aorta
C — Vena cava
D — Pulmonary vein
Q3
1. Blood flows into the atria from the vena cava and pulmonary veins.
2. The atria contract pushing blood into the ventricles.
3. The ventricles contract.
4. Blood flows into the pulmonary arteries and aorta.
5. The cycle starts again as blood flows into the atria.

Page 84 — The Heart and Heart Disease

Q1 a) b) c)

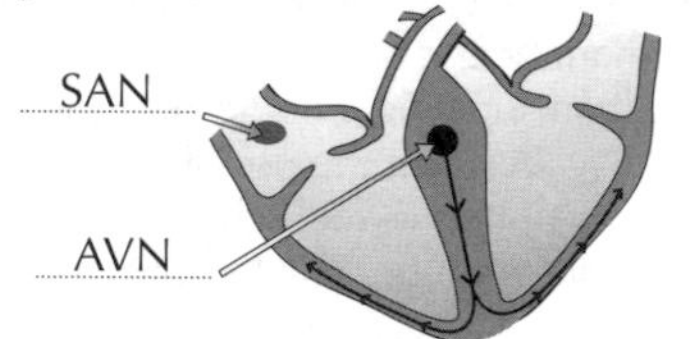

Q2 a) electrocardiogram
b) The electrical activity of the heart.
c) A — atria contract
B — ventricles contract
C — ventricles relax
D — one heartbeat / cardiac cycle
d) The peaks are closer together. / The heartbeat is faster.
e) Echocardiogram / ultrasound scan.
Q3 a) Any five of, e.g. drinking alcohol / smoking / using drugs / stress / poor diet / lack of exercise.
b) Too much saturated fat raises blood cholesterol levels. Cholesterol blocks blood vessels causing heart disease.

Page 85 — Blood

Q1 a) i) Platelets in the blood clump together to 'plug' the damaged area. The platelets are held together by a mesh of protein/fibrin.
ii) Any one of, e.g. fibrin / clotting factors
b) vitamin K
c) green vegetables
d) Alcohol slows blood clotting.
Q2 strokes/DVT, DVT/strokes, warfarin/heparin/aspirin (in any order), haemophilia, clotting factor
Q3 a) During surgery / after an accident / when you've lost lots of blood.

Section Ten — Organs and Systems 2

b) **i)** no
ii) yes
iii) yes
iv) no
c) O
d) Group A blood contains anti-B antibodies. If it came into contact with group B antigens it would cause severe clotting.

Page 86-87 — Waste Disposal — The Kidneys

Q1 E, B, D, C, A
Q2 a) **i)** ions, water, sugar, urea
ii) ions, water, sugar
iii) ions (excess), water (excess), urea
b) Active transport is used to absorb ions and sugar. Water moves by osmosis.
c) **i)** protein and blood cells
ii) They are too big to fit through the membrane.
Q3 Subject 2 might have kidney damage because they have glucose and a lot of protein in their urine.
Q4

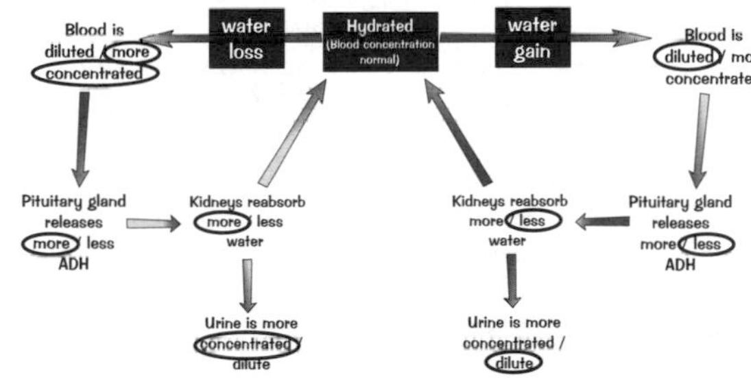

Q5
1. A needle is inserted into a blood vessel in the patient's arm to remove blood.
2. The patient's blood flows into the dialysis machine and between partially permeable membranes.
3. Excess water, ions and wastes are filtered out of the blood and pass into the dialysis fluid.
4. Blood is returned to the patient's body using a vein in their arm.

Q6 a) **i)** proteins and red blood cells
ii) They are too big to fit through the membrane, so they stay in the blood.
b) urea
c) It's the same. This stops glucose diffusing out of the bloodstream and into the fluid, so it's not lost from the patient's body.

Page 88 — Organ Replacements and Donation

Q1 a) They must be fit and healthy, over 18 years old, and a good tissue match.
b) E.g. kidney / piece of liver / bone marrow.
Q2 a) Join the NHS organ donor register.
b) family members
c) E.g. some people think receiving organs is wrong because whether someone lives or dies should be decided by God alone. / Some people think that giving organs is wrong because a person's body should be buried intact.
Q3 a) Heart-lung machine — Oxygenates and pumps the blood
Ventilator — Delivers air to the lungs
Kidney dialysis machine — Filters the blood
b) Any one of, e.g. during or after an operation / when on a waiting list for an organ transplant.
Q4 surgery, success, organ, heart, kidney, age, skill, rejection, drugs

Page 89 — Bones and Cartilage

Q1 a) Supports the body, allows movement and protects vital organs.
b) Any two of, e.g. can grow with the body. / Allows greater flexibility. / Allows easier muscle attachment.
Q2 a) A — cartilage
B — compact bone
C — marrow cavity
b) Protects the end of the bone / stops bones rubbing together at joints.
c) blood cells
Q3 grow, repair, infected, cartilage, calcium/phosphorus, phosphorus/calcium, ossification, cartilage, X-ray, bone, cartilage
Q4 a) Their bones are softer and more brittle, partly due to loss of calcium.
b) If you move someone with a fracture the broken bone may injure nearby tissue.

Page 90 — Joints and Muscles

Q1 hinge, knee/elbow, elbow/knee, one, ball, socket, hip/shoulder, shoulder/hip, rotate
Q2 Any two of, e.g. they don't last forever. / Blood clots are more common. / Hip dislocation is more common. / Risk of infection after surgery. / Inflammation and pain in the surrounding tissue due to the body's reaction to the joint material. / Legs may end up being different lengths.
Q3 Cartilage — Acts as a shock absorber
Synovial membrane — Produces synovial fluid
Ligaments — Hold the bones together
Synovial fluid — Lubricates the joint
Q4 a) A — humerus B — biceps
C — radius D — ulna
E — triceps
b) elbow
c) B
d) E

Section Eleven — Microorganisms and Biotechnology

Section Eleven — Microorganisms and Biotechnology

Page 91 — Bacteria

Q1 a) A = DNA
B = cell wall

b) Any two of, e.g. animal cells do not have cell walls / Animal cell DNA is contained within a nucleus / Animal cells have mitochondria but bacteria don't.

c) Any one of, e.g. mitochondria / chloroplast / vacuole / nucleus

Q2 DNA — genetic material found in the cytoplasm
Flagellum — helps the cell to move
Cell wall — helps to stop the cell from bursting

Q3 a) Clones are genetically identical organisms.

b) The bacterial cell splits into two identical cells.

Q4 Bacteria reproduce more rapidly when they are warm. If food is stored in a fridge, the cold temperature slows down the bacteria's reproduction rate and so the food won't go off as quickly.

Page 92 — Harmful Microorganisms

Q1 Airborne microbes can be breathed in through the nose or mouth, e.g. the influenza virus.
Microbes can enter the body through the mouth in contaminated food or water, e.g. cholera.
Cuts, insect bites and infected needles can introduce microbes through the skin, e.g. malaria / HIV.
Sexual contact can allow microbes into the body, e.g. HIV.

Q2 Once the microbe has entered the body it reproduces rapidly. The microorganisms begin to produce toxins which damage the cells and tissues. The toxins lead to symptoms of the disease developing, and the immune system's reaction can lead to more symptoms such as swelling or fever.

Q3 E.g. developing countries are poorer than developed countries and so are less likely to be able to afford good sanitation and public health measures. There could be a lack of clean water or an ineffective sewage system, both of which can lead to infections spreading. Medical treatment may be insufficient and people may be less educated about hygiene or about how diseases are spread.

Q4 a) E.g. it is difficult for aid workers/health services to reach people in need.

b) E.g. can lead to contamination of drinking water with microbes.

c) E.g. food can't be stored safely and may spoil rapidly increasing the chance of food poisoning.

Pages 93-94 — Microorganisms and Food

Q1 pasteurised, cooled, bacteria, incubated, ferment, lactic acid, clot, flavours

Q2
1. soy beans and roasted wheat are mixed
2. fermentation by *Aspergillus*
3. fermentation by yeast
4. fermentation by *Lactobacillus*
5. filtering
6. pasteurisation

Q3 a) Carbohydrates such as oligosaccharides. / A food source for 'good' bacteria.

b) They don't have the right enzymes.

c) Any two of, e.g. leeks / onions / oats

Q4 a) *Acetobacter* — vitamin C supplement

b) *Aspergillus niger* — citric acid

c) *Corynebacterium glutamicum* — monosodium glutamate

d) *Saccharomyces cerevisiae* — low-calorie sweetener.

Q5 a) It was a control group, to show how/if cholesterol levels changed without stanol esters.

b) To improve the reliability of the results / to lessen the impact of any unusual results / to improve the accuracy of the average.

c) For comparison — to see how it changes during the experiment. / To give a baseline reading.

d) High blood cholesterol increases the risk of heart disease and stroke.

e) Bacteria are used to convert fats called sterols, found in plants like the soya bean, into stanol esters.

Q6 a) Chymosin — clotting agent

b) Fructose — sweetener

c) MSG — flavouring

d) Vitamin C — preservative

Q7 a) citric acid

b) soy sauce, citric acid, fructose

c) soy sauce, stanols, yoghurt

d) carrageenan

Page 95 — Yeast

Q1 a)

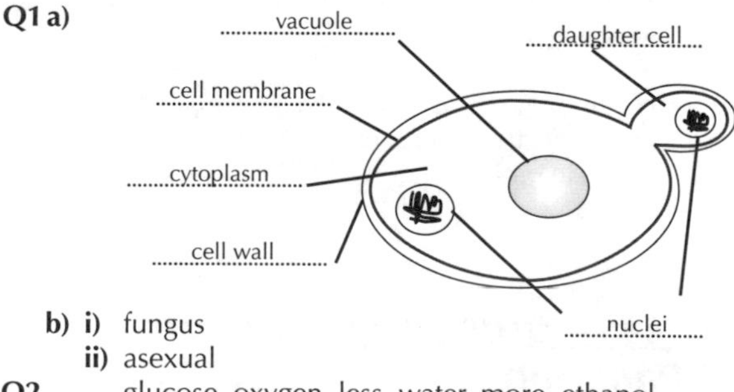

b) i) fungus

ii) asexual

Q2 glucose, oxygen, less, water, more, ethanol

Q3 A — Amount of glucose
B — pH
C — Temperature
D — Level of toxins

Q4 a) Because bacteria would feed on the sugar and reproduce rapidly, removing all the oxygen from the water and causing the death of other water life.

b) Yeast could be used to remove the sugar from the waste water.

Section Eleven — Microorganisms and Biotechnology

Page 96 — Brewing

Q1 a) 1. Barley grains are soaked and germinated.
2. Barley grains are heated and dried.
3. Malted grains are mashed with water.
4. Hops are added.
5. Yeast is added.

b) i) Hops are added to give the beer its bitter flavour.
ii) Yeast is added to ferment the sugar into alcohol.
c) The starch in the grains is broken down into sugar.

Q2 a) The grapes are mashed and water added. Yeast is added and the mixture incubated in a fermenting vessel. The yeast ferments the sugar in the grapes into alcohol. The wine produced is drawn off through a tap and clarifying agents may be added.
b) Beer is pasteurised to kill any remaining yeast and stop fermentation. Otherwise it could spoil if kept in the wrong conditions. Wine isn't pasteurised because it improves the taste if it goes on slowly fermenting.

Q3 a) The fermented liquid is heated to 78 °C to boil the alcohol. The alcohol vapour rises, leaving the water behind, and travels through a cooled tube where it condenses and runs into a collecting vessel.
b) Any two of, e.g. whisky — fermented malted barley / Vodka — fermented potatoes / Rum — fermented sugar cane / Brandy — wine (fermented grapes).

Pages 97-98 — Fuels From Microorganisms

Q1 generator, batch, waste, fermented, heating, turbine

Q2 a) 1. Starch is extracted from a plant.
2. Carbohydrase is added.
3. Yeast is added.
4. The mixture is distilled.
5. The product is mixed with petrol.

b) i) Converting the starch to sugar (which the yeast can then ferment).
ii) Makes the ethanol more concentrated and removes other unwanted substances.
iii) Yeast will ferment the sugar to ethanol.

Q3 a) Methane and carbon dioxide.
b) Any two of, e.g. human sewage / animal dung / kitchen waste (e.g. vegetable peelings) / agricultural/plant waste (e.g. fallen leaves) / sludge waste from factories (e.g. sugar factories)
c) E.g. it's 'carbon neutral'. / Less effect on global warming. / Removes and uses waste. / Produces less of the gases that cause acid rain.

Q4 a) batch
b) continuous
c) all the time
d) continuous

Q5 a) i) So that the people don't have to live near the smell.
ii) It will be easier to transport animal dung to the generator.
iii) To keep it warm so the microorganisms can work effectively.
b) Any two of, e.g. it provides cheap fuel / it disposes of their waste / it improves the properties of the dung as a fertiliser / there's less disease and pollution from the waste / there's no need to spend hours collecting wood for fuel.

Q6 a) Plants obtain energy from the sun. Some of this energy is then passed to animals when they eat the plants. Animal and plant waste is fermented to produce biogas.
b) It can be burnt and used to power an electricity generator.
c) i) It does produce CO_2, but it's derived from recent photosynthesis, which used up the same amount of CO_2, so there's none given out overall.
ii) Any one of, e.g. coal / oil / natural gas / fossil fuel.

Pages 99-100 — Enzymes in Action

Q1 a) An enzyme is a protein that speeds up/catalyses a chemical reaction in a living organism.
b) Any three of, e.g. in biological washing powders / in stain removers / to make low-calorie sweetener / to make cheese / to extract fruit juice / in biomedical industry to test blood/urine sugar levels.

Q2 a) Proteins don't dissolve in water so can't be washed out. The protease breaks down the proteins in stains like blood, egg etc. to amino acids. Amino acids can be washed out, removing the stain.
b) So the washing powder can be used at high temperatures without damaging the enzymes. A high temperature is more effective at removing other types of stain.

Q3 a) Used to clot milk in cheese production.
b) Used to change sucrose into glucose and fructose. which is used to make low-calorie food sweeter.
c) Used to break down pectin in the cell walls of fruit to help extract the juices for fruit juice.

Q4 a) Any three of, e.g. more convenient / faster / less equipment required / specific to glucose / easier to determine the concentration because of the different colours produced.
b) E.g. used to determine blood/urine sugar levels in order to detect/treat/monitor diabetes.

Q5 a) E.g. they can be attached to an insoluble material such as silica gel/fibres of collagen/cellulose.
OR they can be encapsulated in alginate beads.
b) The enzyme does not contaminate the product.
The enzyme can easily be washed and reused.
The enzyme is more stable/less likely to denature.

Q6 a) Some people don't produce the enzyme lactase so they can't break down lactose. This causes digestive problems if they drink milk.
b) The milk is run through a column of immobilised lactase enzymes, which break down the lactose in the milk. Lactose-free milk then emerges from the bottom of the column.

Page 101 — Genetically Modifying Plants

Q1 C, E, A, D, B

Q2 a) They're used as a control, to show how plants that could not have been affected by the GM wheat respond to spraying.
b) They could have used larger groups of plants / more study sites.

Section Twelve — Behaviour in Humans and Other Animals

c) Their conclusion is correct. A difference of 2 plants out of 100 could be due to chance.
d) No. Just because the genes didn't spread in this experiment doesn't mean that they won't in the future. A lot more evidence would need to be gathered before you could be reasonably certain that the GM crop posed no threat.
e) The wild grasses could grow over farmers' fields and decrease crop yields — the farmer would not be able to remove them using the herbicides he would normally use.

Page 102 — Developing New Treatments

Q1 genomics, medicine, predispose, prevent, design
Q2 a) active ingredient
b) malaria
Q3 a) It has increased every year, and more rapidly in recent years.
b) Any two of, e.g. drugs are becoming more expensive. / New drugs are becoming available. / People are needing/demanding more medical attention. / The country is getting richer and can afford to spend more on drugs.
c) E.g. Country B might not be able to afford the new drugs available / they may have prevented more cases of disease.
d) If other companies were allowed to copy new drugs the price could be lowered. This would mean country B could buy drugs that were too expensive before.
e) Companies would have no incentive to do the research needed to develop new drugs, so no new drugs would be developed.

Section Twelve — Behaviour in Humans and Other Animals

Pages 103-104 — Instinctive and Learned Behaviour

Q1 learned, genes, light
Q2 Playing football — learned
Salivating — instinctive
Language — learned
Sneezing — instinctive
Q3 a) Any food preference (e.g. robins preferring cheese). These preferences are found only in certain species, so are likely to be determined by genes.
b) The crow hanging from the string to obtain the nuts — it developed over a period of time after the bird observed the great tits.
c) It might not develop the proper song for its species.
Q4 a) An animal learns not to respond to a stimulus that is neither beneficial nor harmful to it.
b) By ignoring neutral stimuli, animals spend their time and energy more efficiently.
Q5 A Skinner box is a small cage where an animal has a choice of buttons to press. When it presses one particular button it is rewarded with food. Animals use a system of trial and error to learn which button to press, so the box can be used to study how different species learn.
Q6 a) Classical conditioning is passive learning, where an association is formed between a 'neutral' stimulus and one that naturally brings about a response. It's automatic and reinforced by repetition. Operant conditioning is active learning where an association is formed between an action and a reward or punishment by trial and error.
b) i) Any sensible suggestion, e.g. dogs learn to associate a bell ringing with food and then salivate at the ring of a bell (Pavlov's experiment).
ii) Any sensible suggestion, e.g. rats learning to obtain a food reward by pressing a button/lever (Skinner's experiment).
Q7 a) E.g. food treat or praise when the dog behaves correctly / verbal reprimand or choke chain when the dog doesn't stop and wait.
b) Any one of, e.g. some people think that conditioning involving punishment is cruel / stresses the animal / doesn't work any better than conditioning involving reward.

Pages 105-106 — Social Behaviour and Communication

Q1 Any three of, e.g. to warn others about predators or other dangers. / To inform others about the presence of food. / To help keep a group together. / To allow predators to coordinate their hunting. / To communicate mood and so avoid unnecessary fighting. / So baby animals can communicate their needs to their parents. / To inform potential mates that they are ready to breed.
Q2 a) To show that she is ready for mating / to attract a mate.
b) To communicate where a food source is.
c) To show that it is submissive / to avoid a fight.
Q3 a) E.g. to attract / impress potential mates.
b) E.g. the feathers are likely to attract the attention of predators. / They make it difficult to fly and escape from predators.
c) Females choose males, so they don't need to be brightly coloured to impress a mate. / Females incubate eggs and care for young, so need to be camouflaged to avoid attracting predators.
Q4 a) Any three sensible suggestions, e.g. body language, facial expressions, any specific examples of these (smiling, grimacing, frowning, etc.), nodding, shaking the head, written or signed language.
b) Verbal communication can provide more specific / detailed information.
Q5 a) With their green-brown colour, it is difficult for them to see one another in the foliage of the trees.
b) Since the birds look very similar, their songs must be different so that they attract members of the correct species.

Section Twelve — Behaviour in Humans and Other Animals

Q6 a) Language is more complex than birdsong — it can be used to express knowledge of past events, emotions, abstract ideas, etc. / Language is symbolic, with words being used to represent objects or ideas, and birdsong is not.

b) No, because facial expressions are species-specific — a frown could mean something different to pandas.

Q7 It suggests that the dog doesn't realise it is seeing itself — therefore it isn't self-aware. Humans are different in that they're conscious of their own existence.

Q8 a) Some people define self-awareness as simply being aware of your own existence. Others think that it is also about consciousness (awareness of your own behaviour and feelings) and accountability (awareness of the possible consequences of your behaviour).

b) We don't know what an animal is thinking (or if they 'think' at all).

Pages 107-108 — Feeding Behaviours

Q1 herbivores, herds, predators, amino acids

Q2 a) herbivores

b) carnivores

c) carnivores

d) herbivores

e) carnivores

Q3 a) Because a single wolf would not be able to kill a large animal like a reindeer / could be harmed by the reindeer.

b) The reindeer are more likely to spot predators early with so many pairs of ears and eyes grouped together.
Any one individual reindeer is less likely to be picked off, as it's harder for the wolf to get access to the reindeer if they are closely packed together.

c) They eat all the available food in an area so need to find new feeding areas.

d) Smaller prey are easier to catch. / Smaller prey would not provide enough food for the pack to share.

Q4 a) It is the red spot that stimulates the begging response. The pointed red stick, which looked least like a bird's head, produced the biggest response, showing that the shape of the head and the colour of the bill were unimportant. The head model that lacked a red spot produced the least number of pecks.

b) Instinctive — young birds do it soon after hatching / there is no time to learn it.

c) Different coloured models with different sized mouths could be presented to parent / adult birds to see which stimulates regurgitation.

Q5 a) Any two of, e.g. chimps using twigs to dig up edible roots/get honey from beehives/get ants from their holes. / Chimps using leaves to wipe blood, fruit etc. from their fur. / Woodpecker finch using cactus spines to lever grubs out of tree bark. / Monkeys using pieces of wood as spoons.

b) No, a tool is an object used as an extension of the body — the rock isn't picked up by the vulture so isn't used as an extension of its body.

Pages 109-110 — Reproductive Behaviours

Q1 mandrill monkey — display brightly coloured parts of body
red deer — display aggression to other males
frog — mating call
moth — pheromone

Q2 a) Staying with just one mate.

b) A group of females that all mate with a single male.

c) Behaviours seen in many animals before mating.

Q3 a) i) E.g. they get more food (so grow more quickly). / They are defended better against predators. / They are more likely to survive.

ii) Each parent spends less time and energy in looking after young, so can spend more time in feeding themselves. / More young are likely to survive into adulthood and pass the parental genes on to the next generation.

b) It doesn't take both parents to raise young to adulthood because there are fewer dangers, so the male has more to gain by mating with many females and passing on his genes than by helping to raise one set of chicks.

Q4 a) Females of most species invest more in each offspring than males, which usually breed with several females and give less time and energy to producing and raising the young. The female needs to select a strong mate to give her young the best genes and so the best chance of survival, and the male must prove he is worthy. However, in seahorses it is the male that invests more time and energy, so the roles are reversed and females must compete to win the right to lay eggs in his pouch.

b) Because the resultant offspring would be infertile (if any were born at all) and all her efforts to pass on her genes would be wasted.

Q5 a) Care: any kind of mammal, most birds, crocodiles, some fishes (e.g. seahorse, cichlid fish, etc.).
Don't care: most invertebrates, reptiles, amphibians and most fishes.

b) Any three of, e.g. parents fend off predators / keep young warm / feed young / teach young skills for survival / warn young of danger / find or build shelters or nests.

Page 111 — Living in Soil

Q1 a) Waterlogged soil contains less oxygen, so the earthworms were coming up to the surface for air.

b) Fewer dead leaves / organic matter would be buried so there would be fewer nutrients in the soil. There would be less air in the soil, less mixing of the different layers and it would become more acidic.

Q2 a) Saprophytic bacteria decompose dead matter and release ammonium compounds. Some nitrifying bacteria (e.g. *Nitrosomonas*) convert the ammonium compounds into nitrites and then other bacteria (e.g. *Nitrobacter*) convert the nitrites into nitrates. These are then taken up by plants.

b) E.g. The farmer could add a nitrate fertiliser / add more dead material / add more bacteria / grow legume plants (which have roots containing the bacteria).

Section Twelve — Behaviour in Humans and Other Animals

Q3 a) Nitrogen from dead leaves goes straight into the trees instead of into the soil.
b) The roots get their nutrients from dead leaves which are found near the surface of the soil.

Pages 112-113 — Living in Water

Q1 Advantages — any two of, plentiful supply of water / less variation in temperature / the water provides support / waste disposal is easier.
Disadvantages — more energy needed to move through water / effects of osmosis could damage cells so water intake must be carefully regulated.
Q2 The robber crab, because there is more variation in temperature on the land where this crab lives.
Q3 Water provides support for aquatic plants, so there is no need for them to be supported by woody parts.
Q4 a) The solute concentration in amoeba cells is greater than the solute concentration in the water. This means that water moves into the amoebas by osmosis.
b) Water taken in by osmosis is collected in the contractile vacuole which then contracts to empty the water outside the cell.
c) The solute concentration in the cell is the same as or lower than that outside, so no water enters by osmosis and there's no need to get rid of excess water.
Q5 a) Phytoplankton are microscopic water plants/algae. Zooplankton are microscopic water animals.
b) Phytoplankton can photosynthesise and are important producers in aquatic food webs.
Q6 a) In the summer there is more light and a higher temperature, allowing a faster rate of photosynthesis and faster growth.
b) More phytoplankton means there is more food for zooplankton, so the zooplankton increase in number.
Q7 a) i) Through the lungs and the skin.
ii) The skin needs to be moist to help gaseous exchange through it. This means they have to live in moist environments. (Their skin is not waterproof, so they would lose too much water in a dry environment.)
b) i) Gills. They are highly folded to give them a big surface area for gaseous exchange.
ii) Water supports the gills by keeping the folds separated from one another. Out of water, the gills stick together and the fish suffocates.

Page 114 — Human Evolution and Development

Q1 a) Tool use has enabled humans to modify their environment — to produce food by farming, to clear land and build, to travel further, etc.
b) Living in societies enables cooperation between individuals — knowledge and work can be shared and so more can be achieved.
Q2 a) People don't farm to get food, but hunt for animals and gather vegetables, fruit, etc. found growing wild.
b) E.g. tools could be used for hunting (e.g. spears), or for constructing means of carrying food (e.g. baskets).
c) Less time needs to be spent hunting and gathering if plants and animals are farmed near home. / A constant supply of food is available for most of the year.
Q3 a) E.g. dogs could be used to help hunt other animals / for companionship / for protection / for food.
b) Horses: A, B, C, D, E
Cattle: A, B, C, E
c) Any one of, e.g.
Farm animals — to produce more/better quality meat/milk/wool, etc.
Horses — for speed/strength.
Guard dogs — to be more aggressive.
Most animals — to have a better temperament / to be easier to handle.

Page 115 — Human Behaviour Towards Animals

Q1 a) Any one of, e.g. sheep, goats, rabbits.
b) Any one of, e.g. deer, game birds, foxes.
c) Any one of, e.g. horses, dogs.
Q2 For: e.g. zoos can be used to save endangered species from extinction. They have been used to help restock wild populations. They are important for educating the public.
Against: e.g. it's unnatural and cruel to keep animals in confinement. Resources should be focused on protecting wild populations.
Q3 a) To ensure that harmful effects are discovered before the drugs are given to humans.
b) Any one of, the drugs could cause the animals pain and suffering. / They may be stressed by the confinement or treatment.
c) Any one of, e.g. they could be genetically engineered so that they produce drugs in their milk/ used to provide antibodies for use in vaccines/used to provide organs for use in transplants/to study the way diseases are contracted, develop and affect the body.
Q4 a) i) Any one of, e.g. the animals have to be kept enclosed/transported often/made to perform, which could be stressful. / Their conditions are likely to be far more unnatural than those found even in zoos.
ii) Many circus owners claim that their animals are well-treated and enjoy performing.
b) i) Animals are sometimes kept in cramped conditions where they can't move about/socialise which makes them more susceptible to infectious disease.
ii) Intensive farming produces more food for humans more cheaply, making it more affordable.

BHW43